D1264248

CAMBRIDGE MONOGRAPHS ON PHYSICS

THE CONCEPTS OF CLASSICAL THERMODYNAMICS

THE CONCEPTS OF CLASSICAL THERMODYNAMICS

BY

H. A. BUCHDAHL

Professor of Theoretical Physics,
School of General Studies,
Australian National University

CAMBRIDGE
AT THE UNIVERSITY PRESS
1966

PUBLISHED BY
THE SYNDICS OF THE CAMBRIDGE UNIVERSITY PRESS
Bentley House, 200 Euston Road, London, N.W. 1
American Branch: 32 East 57th Street, New York, N.Y. 10022
West African Office: P.M.B. 5181, Ibadan, Nigeria

Printed in Great Britain at the University Printing House, Cambridge
(Brooke Crutchley, University Printer)

LIBRARY OF CONGRESS CATALOGUE
CARD NUMBER: 66-10176

CONTENTS

CHAPTER 5

The Second Law (I)

CHAPTER 6

The Second Law (II)

CHAPTER 7

The Third Law

CHAPTER 8

Potentials, Constitutive Coordinates, and Conditions of Equilibrium

CHAPTER 9

Miscellaneous Topics

CHAPTER 10

Applications (I)

CHAPTER 11

Applications (II)

PREFACE

There exist today many excellent treatises on classical thermodynamics such as those of Guggenheim, Landsberg, Pippard, and Wilson, to mention but a few. It is not my purpose to supplant them, even were I competent to do so. On the contrary, this small volume is intended merely to present the basic ideas of classical thermodynamics on a purely phenomenological level, against a general background of physical theory. This is done in a way which should give the reader a feeling of easy familiarity with the laws to be introduced and the concepts to be defined. It is just this feeling which the student so often lacks after a first study of the subject. In a sense, therefore, my aim is very circumscribed. There will be no attempt at completeness or generality, and all manner of simplifying assumptions will be introduced from time to time. This procedure is, I think, likely to enhance the reader's chances of getting a basic understanding of the subject, and any required generalizations can be undertaken at a later stage.

My general approach tries to couple simplicity with reasonable sophistication. It should be clearly understood from the outset that, superficial appearances to the contrary, this is *not* an axiomatic treatment of the subject. Undeniably useful though an axiomatic treatment of a physical theory may be, it too easily gives the appearance of a logical exercise, and were this to be the case here the whole purpose of this work would be vitiated. Moreover, to carry through a valid axiomatic development is likely to require at times assumptions so restrictive as to put its relevance to *physical* theory in doubt. This point of principle seems to be sometimes overlooked. At any rate, frankly pragmatic though the character of the present approach may be on occasion, it is constantly governed by the main aims of simplicity and clarity, however much I may in fact have failed to achieve these.

Though this volume is reasonably self-contained, it might well be read alongside one of the standard treatises already referred to. I adhere strictly to the phenomenological point of view, not least in order to counter the widely held opinion that one can gain an understanding of the concept of entropy, for instance, only on a statistical basis. In the first place, it might be reflected that the theoretical foundations of statistical thermodynamics are difficult to a degree which makes it doubtful whether the belief in question can be sustained. Be that as it may, it involves an obvious logical difficulty: a statement of the kind 'the results of statistical thermodynamics must not lead to contradictions with the laws of phenomenological thermodynamics' becomes meaningless if the latter can be 'understood' only on the basis of the former: to grant that it is meaningful is to grant that the conceptual frameworks of both theories are *separately* well defined.

After some general remarks concerning physical theories as such, the subject is developed in a manner which does not place undue verbal emphasis on terms characteristic of the realm of engineering. Moreover, the introduction of ideal gases into the basic parts of the theory is avoided, for there is a didactic weakness inherent in an early appeal to the properties of a class of substances whose existence in nature is later denied by one of the principal laws of thermodynamics. The four main laws are stated in their traditional order even though, from a pedagogic viewpoint, there are advantages in stating the second before the first. With regard to the former, Carathéodory's formulation has been adopted, since this allows one to achieve a clearer separation between the mathematical and the physical content of the theory, even if, contrary to a widely held view, its supposed greater logical economy appears to be somewhat illusory. The concept of entropy is introduced in various alternative ways, with a certain emphasis on the idea of an ordering of states. Some redundancy is of little account: I have not aimed at the greatest possible brevity. Indeed, as a result of closely scrutinizing points which are often scarcely examined at all, this exposition may appear somewhat old-fashioned in its discursiveness. It may not be out of place to emphasize that it is intended to be of an elementary character, though here and there it may reveal itself as being a little more sophisticated than most elementary treatments of the subject; an occasional reference to an unfamiliar branch of physics may simply be left unread.

The systematic development of the general theory has been kept unencumbered by discussions of specific applications. These have been relegated to the last two chapters, where they are presented in an orderly fashion on the basis of a classification which assigns to any particular application a certain character. It will be appreciated that the material of these chapters is mainly intended to serve a methodological purpose. Bearing in mind at the same time that this work will mostly be used as collateral reading, the absence of numerical examples should occasion no surprise.

My warmest thanks go to Professor Arthur Hambly and to Dr Daniel Greenberg for reading the whole manuscript, and to Mr Malcolm Urquhart for reading the first three chapters. Their advice has been of great value to me. The responsibility for any errors and omissions which remain is of course entirely mine. To Mrs Myrene Hickey I wish to express my appreciation of the excellence and patience she displayed in typing the manuscript.

Finally I should like to thank the Cambridge University Press most warmly for their unfailing care and courtesy.

Canberra H.A.B.
6 June 1965

CHAPTER I

INTRODUCTION

1. The nature of physical theory and related notions

(*a*) Classical thermodynamics is a physical theory which deals with certain very general features of the overall behaviour of macroscopic systems on the basis of four principal laws and some special concepts related to these. A broad statement of this kind is not very meaningful unless some explanation be provided of the various terms contained in it. Therefore, the first task to be accomplished is to undertake a short survey of the meanings which might be attached to terms such as 'theory', 'law', 'physical quantity', and the like, in particular in as far as they relate to the subject in hand.

(*b*) What then is a physical theory? In its widest sense it may be taken to be a systematic scheme of statements held as an explanation or description of some set of natural phenomena which either have been observed in the past or are expected to be observed in the future under certain conditions. The assignment of such conditions requires the description of certain 'things', that is, some set of external objects which are the carriers of certain properties of interest. In any particular case one refers to such a set of objects as the 'physical system'. When it is, at least in principle, accessible to direct perception it is 'macroscopic'; so that one might therefore simply point to an appropriate collection of objects and declare it to be the system under investigation. Let this have been done.

(*c*) Further progress is, to begin with, of an empirical kind. A set of prescribed experimental procedures is carried out: the final outcome of which is to be in every case a number. For example, given a rectangular table T, one such procedure might consist in determining how many unit measuring rods (which are here taken for granted) must be laid end to end in the usual way along one edge E to reach from one of its extremities to the other. The result of this operation is a number, and it must be required that upon repetition under the same circumstances the number obtained will be the same, at any rate to within experimental inaccuracies, a complication which will here be disregarded. This number, obtained in the manner just described, is the value of a *physical quantity*, usually called the 'length' of E. (It may be noted in passing that the existence of such a fixed number underlies the previous remark that physical systems are carriers of properties.) One step removed from the kind of operations envisaged hitherto are those in which calculations take part; these, however, for the present being supposed to be merely essentially arithmetic operations on the numbers arrived at in the manner described earlier. To stress the idea of operation here also, one occasionally refers to such calculations as 'paper and pencil operations'. For example, one might find the lengths of two adjoining edges of the table and calculate the product of these two numbers: the result is the value of another physical quantity, called the 'area' of T. Quite generally the following guiding principle will henceforth be adopted:

> *A physical quantity is defined by a consistent set of unambiguously prescribed operations.* (1.1)

To clarify the distinction between a physical quantity Q and the value q of this quantity it suffices to choose Q as 'length'. The length of any object is that of its properties, or characteristics, the investigation or determination of which proceeds according to the relevant procedure implicit in (1.1) (in this case with measuring rods laid end to end); that such a relevant procedure is meaningful, i.e. exists, being taken for granted in this context. The value of the length is the number which results from the procedure in any particular case.

It is true that all manner of objections may be raised against the point of view here adopted. For instance, it apparently leads to a proliferation of physical quantities. Thus, in the example of the table, one might measure the time in seconds which it takes for a ray of light to travel along E and back again, and multiply the number so obtained by $1\cdot5 \times 10^{10}$. Force of habit might incline one to refer to the resulting number also simply as the length of E. This, however, would imply an inconsistency, for the new set of operations is entirely different from the first. At best, one could speak of the 'rod length' and the 'mirror length' of E, so making the duplication of the original physical quantity explicit. Appearances notwithstanding, this state of affairs is as it should be: the fact that experimentally the two lengths turn out to be the same is to be regarded as a physical law, and not as something obvious *a priori*. Indeed, one has simply a restatement of some of the laws of electromagnetism. It seems, then, that the particular objection under discussion is without substance; and although more intractable objections might be raised I shall adhere to the operationalist principle (1.1) laid down above, for even if it be used only heuristically, it is of inestimable help in charting a clear course through a theory and avoiding the appearance of tautologies, void definitions, circular arguments, and the like.

(*d*) In addition to physical quantities one customarily speaks also of physical 'concepts'. A concept being a general notion or idea, one enters here to some extent into the realm of intuition, with all its attendant difficulties. It is probably fair to say that in speaking of the 'concept of length', for instance, one presupposes an intuitive idea of some absolute property of objects which goes beyond the mere operational notion of length as discussed previously. Indeed, it seems at times to be implied or assumed that this intuitive length can be 'measured in different ways', e.g. by means of rods, or light and mirrors, or by determining apparent diameters and the like; and that the choice of one or the other is merely one of convenience, the result in every case being of necessity the same. In the light of the position taken here, however, one is, on the contrary, confronted with *different* physical quantities, as has already been remarked: and coincidence of their values in any particular instance must in general be regarded as an empirical result, rather than as something to be accepted *a priori*. That these are not idle remarks will be realized by anyone who has even a fleeting acquaintance with cosmological theory. There one has all manner of 'distances', for example luminosity distance, parallax distance, and distance by apparent diameters; and the values obtained according to the appropriate prescription in each case will in general be different for a sufficiently remote object.

Nevertheless, use of the term 'physical concept' is often appropriate. This is the case, for instance, when one wishes to refer to some class of physical quantities which under certain conditions are known empirically to have the same values. This use of the term is coherent with the remarks just made about distance. Again, it is useful in overcoming verbal difficulties engendered by the arbitrariness of scales and units inherent in the definitions of physical quantities. Thus, the definition of length involved a 'unit rod', but it might well be convenient to use one kind of unit rod for one purpose but another kind for a different purpose. To some extent this again gives rise to a proliferation of physical quantities, though perhaps a less serious one. At any rate, it is convenient to have a term which describes the notion of rod-length in general: and this is the *concept* of rod-length. In other words a physical concept, as understood in this context, is the general notion of a set of corresponding physical quantities defined by the same generic description but differing in the adoption of units or scales implicit in them. Evidently, the terms quantity and concept may occasionally be used interchangeably; but as soon as reference is made to 'values' one is restricted to the use of quantity alone.

(*e*) Given a physical system, one associates with it a set of physical quantities in the manner outlined above by means of largely empirical procedures. The kind of quantities which can be selected, or which one does in fact select, are characteristic not only of the system as such, but also of the particular aspects of its behaviour which one wishes to investigate. The choice of such quantities having been decided upon at the outset, the theoretical part of the investigation consists in the search for relations between these quantities, that is, between the values which they take under varying conditions. Thus, having gone through say n generically different routines, resulting in the values $q_1, \ldots, q_n$ of n different physical quantities $Q_1, \ldots Q_n$, one might find that the nth of these is in every case deducible from those of the remaining $n-1$. One verbalizes this situation by declaring that the nth quantity is such-and-such a function of the other $n-1$, and writes

$$q_n = f(q_1, q_2, \ldots, q_{n-1}). \tag{1.2}$$

Such an expression of an *observed* regularity will in general have the status of a 'physical law', granted that the content of (1.2) does not happen to be trivial. ((1.2) would be trivial if Q_n were defined in terms of $Q_1, \ldots, Q_{n-1}$.) A physical law may of course have to be expressed in a manner different from that just considered. For instance, it might take the form of a differential or integral equation, or, again, it might be in effect a statement concerning the existence of certain relations between sets, and so on. In every case, however, it is a generalization from experience suggested by the examination of a necessarily finite number of experimental findings, and therefore in the first place describes merely what was observed in the past. It goes, however, much further, in that the law is intended to cover also the outcome of experiments yet to be performed, and is therefore endowed with *predictive* power. (Any such empirical law is certainly fallible, for infallibility would imply that the observed relationships are after all the result of logical necessity.) Over and above this, 'assumptions' are often introduced which have their origin in considerations of plausibility and which are, in the first place, not based directly upon experience. Assumption of continuity and differentiability fall under this heading; their validity must then be tested by their empirical consequences. Finally, then, as an alternative to what was said earlier in this section, a physical theory may be regarded as constituted of certain laws expressing inter-relations between members of a set of relevant physical quantities, together with any ancillary assumptions concerning them which may have been introduced.

2. Presupposition of mechanical concepts. Motivation of thermodynamic theory

(*a*) Classical mechanics is the theory of motion of material bodies under the influence of their mutual interactions. In particular all velocities and accelerations in a mechanical system may happen to be zero: one has a state of 'static equilibrium'. Characteristic of this theory are certain concepts such as kinetic and potential energies, pressure, work, elastic forces, and the like. *All these are henceforth to be taken for granted.*

(*b*) The energy of a mechanical system is conserved; that is to say, the work done on a system reappears in the form of kinetic and potential energy, and it is entirely recoverable. This is to be understood as follows. The most cursory investigation reveals that mechanical energy is frequently not conserved, for instance when a fluid undergoing any sort of motion within a rigid enclosure eventually comes to rest. The contradiction implied in this observation is purely verbal. It is true that 'mechanical energy' has failed to be conserved; but this merely means that the physical system in question is not a 'mechanical system'. Traditionally one refers to such non-conservative systems (in which no electromagnetic forces operate and neither forces nor constraints depend explicitly upon the time) as 'dissipative mechanical systems'. However, it is more appropriate in the present context to adhere to the convention that they are not to be described as mechanical systems at all: (i) because then one cannot inadvertently presuppose concepts (such as temperature) yet to be defined; and (ii) because it is precisely the recognition of the existence of such non-conservative systems which gives rise to *thermodynamic theory*, in which certain general conclusions are drawn about their behaviour which go beyond the laws governing the behaviour of (macroscopic) mechanical systems (see also Section 13).

3. Terminal conditions. Thermodynamic systems

Let it be imagined that work is done on some mechanical system, and that the latter is both initially and finally in static equilibrium. Its energy will have increased, and this increase depends only on the relative displacement of its component parts, in short, its 'deformation'. This deformation is defined by the change in the values of certain 'deformative' quantities; and these are determinable by measurements of a *geometrical* kind alone. The same is therefore true also of the changes of energy if one takes into account changes of external potential energy and of the potential energy of elastic strains. (The system being purely mechanical, these strains depend solely upon the stresses responsible for them.) The central point here is this: that the changes of energy brought about by external interactions are fixed entirely by the deformations which it induces.

As adumbrated in the preceding section the situation is otherwise when the system is not purely mechanical. By way of a simple example, consider again a fluid entirely filling a rigid enclosure. Work may be done on the fluid by stirring it. It shall have been at rest initially, and it will be at rest again some time after the stirring has ceased. The geometrical aspect of this system will be the same finally as it was initially, it being understood that the stirring mechanism is not to be regarded as part of the 'system'. Thus, though work has been done on it no deformation has occurred, in contrast to the mechanical situation discussed earlier.

This is not to say that the 'terminal conditions' (i.e. initial and final conditions) of the system are the same in all respects. On the contrary it is a matter of experience that one can associate a further, non-deformative, quantity with it whose terminal values will in general differ from one another. Only when this new quantity is adjoined to the old will a complete characterization of the terminal conditions of the system be possible. In the present situation the volume x_1 of the enclosure might be chosen as the single relevant deformative quantity; the pressure x_2 exerted by the fluid upon its enclosure on the other hand will serve as a non-deformative quantity.

The implications of these remarks are of course intended to go beyond the narrow confines of the special example above. That is to say, the class of *thermodynamic systems* is characterized by the requirement that every terminal condition of such a system K is defined by the values of a finite set of relevant macroscopic quantities, at least one of which must necessarily belong to the nondeformative type (see also Section 10).

Even if K be a physico-chemical system, 'constitutive' quantities such as those describing the amounts of substances present in the various phases are here to be left out of account. Although they are of considerable interest they take part in defining the *internal constitution* of the system, whereas the 'relevant' quantities are external in the sense that they enter directly into any relation describing the interaction of K with its surroundings, granted of course that the total amount of matter within K is kept fixed (see also Section 65).

4. Equilibrium

A moment's reflection will show that the 'terminal conditions' repeatedly referred to above are analogous to the static equilibria of mechanical systems, to the extent that they imply the absence of any macroscopically observable motions of parts of the system (see also Section 76*a*). In a general context, however, the term 'equilibrium' has so far been avoided.

In the first place it should be recalled that a terminal condition is (by definition) static. This certainly seems to be meaningful, for whether a system K is undergoing a change of some kind should surely be observable. Yet a difficulty of time scale, in the colloquial sense of the term, arises here. For it might well be the case that an experimental investigation of a given system does not reveal any measurable changes during the time required to carry out the experiments; but this is not to say that a change might not have been observed had the investigation extended over a greater interval of time. This situation arises, for instance, in systems containing substances whose properties at a given time depend upon their previous history. Some kind of internal conversion of a working substance may merely be inhibited, so that the system may appear to be static under some conditions. Yet the fact that internal conversion may proceed at an enhanced rate when the conditions are varied means that in the first instance the system was not properly static after all.

If the system is in a terminal condition which is properly static it will be said to be in *equilibrium*. Systems which cannot attain equilibrium under appropriate experimental conditions will henceforth be excluded from consideration. Of course, one may sometimes—strictly speaking, perhaps always—have to assume that a given terminal condition is one of equilibrium, even when the possibility of a direct verification of the truth of this assumption is not at hand. In that case subsequent empirical verification or falsification of certain theoretical conclusions, based in part on the assumption in question, must serve instead (see also Section 61*b*).

This pragmatic approach to the notion of equilibrium is probably unavoidable. To begin with it is not very easy to see how the difficulty already raised can be otherwise circumvented on a phenomenological level. Furthermore, the idea of the 'approach to equilibrium' gives rise to much the same kind of intractable problem. Consider the customary phenomenological description of the diffusion of one gas into another by means of the diffusion equation. Its solution would seem to indicate that at any point the concentration of either gas would never reach a value constant in time, though it approaches such a value asymptotically. The settling-down process of a gas initially in turbulent motion presents an analogous situation. Evidently, in the first case, one has to take the view that differences between the actual values of the concentration and the values they approach asymptotically are physically meaningless after a certain time; whilst in the second case the gas may be considered as strictly at rest after a certain finite time ('relaxation time'). Such a view must in any case be adopted as soon as the discontinuous structure of matter is taken at least qualitatively into account.

5. Coordinates

In accordance with previous definitions, the equilibrium of a thermodynamic system K is characterized uniquely by the values of a *finite* set of physical quantities. It may happen that when such a set has been chosen one or more of its members turn out to be redundant; and these may then be excluded. In this way one is left with a 'minimal set'. Of any such minimal set it will be required as a matter of convenience that it contain all the quantities required to define the configuration of K, i.e. those which are purely deformative; when this condition is satisfied one has a 'proper minimal set'. (For example, a possible set of quantities for the system of Section 3 would be x_1, x_2 and $x_3 = x_1^2 x_2$. It is not minimal, and one of these three quantities is to be excluded. This may be either x_2 or $x_1^2 x_2$ but not x_1, if the minimal set is to be proper.) For any given K, the choice of a proper minimal set is by no means unique, whilst the number n of members it contains is a matter for empirical decision.

The n quantities of any particular proper minimal set are called *coordinates* of K. The element of arbitrariness left in the choice of coordinates makes them to some extent analogous to the generalized coordinates of analytical dynamics. However, in the thermodynamic situation, one has the restriction, already imposed, that amongst the coordinates all non-redundant purely deformative quantities must be present: and these will be called *deformation coordinates*. It should be noted that this name is to some extent conventional. For example, if a magnetic substance is present in K one or more 'deformation coordinates' will be required to specify its magnetic moment.

6. States. Transitions

Any particular set of values of the coordinates of K is a *state* of K, and the passage from one state to another is a *transition* of K. With regard to these definitions it should be kept firmly in mind that the notion of 'state' clearly refers only to equilibrium conditions. If the coordinates of K be denoted by $x_1, x_2, \ldots, x_n$ the phrase 'K is in such-and-such a state' means that K is in equilibrium, the x_k having certain fixed, well-defined values. It is not necessary to say (except for the sake of emphasis perhaps) that K is in an 'equilibrium state', for strictly speaking this appears to suggest that there exists the notion of a 'non-equilibrium state'. Recalling the various definitions as set out here, this notion is, however, meaningless, for when K is not in equilibrium some of the coordinates have no well-defined values. (For example, no finite set of numbers, intended to represent 'the pressure', can be assigned to a gas moving turbulently within an enclosure.) Coherently with this, since except under special circumstances a system will not be in equilibrium in the course of a transition, it will generally be in some state only initially and finally, whilst during the transition it will be in no state at all. This terminology may appear slightly idiosyncratic, but it is both convenient and rigorous. In parenthesis, it may be noted, now that 'state' has been defined, that coordinates are perhaps more often referred to as 'variables of state'.

7. Pseudo-static, quasi-static, and reversible transitions

(*a*) A continuous sequence $\mathfrak{C}$ of states $\mathfrak{S}$ of K may be defined by the equations

$$x_k = f_k(t) \quad (k = 1, 2, \ldots, n), \tag{7.1}$$

where $f_k(t)$ is an arbitrary continuous function of a parameter t, $(t' \leqslant t \leqslant t'')$. One can arrange two prescribed states $\mathfrak{S}'(x'_k)$ and $\mathfrak{S}''(x''_k)$ $(k = 1, 2, \ldots, n)$ to be the terminal states of $\mathfrak{C}$ by choosing the f_k to be such that

$$f_k(t') = x'_k, \quad f_k(t'') = x''_k. \tag{7.2}$$

Then, a transition between two states is *pseudo-static* if in its course K goes through a continuous sequence of states. Such a transition may therefore be parametrized after the fashion of (7.1).

Any pseudo-static transition must proceed at an 'infinitesimal' rate, for it is a matter of experience that in a system undergoing a rapid change all manner of phenomena, such as turbulence within working substances, the generation of elastic waves and the like, preclude equilibrium. If a transition is not pseudo-static then it is called *non-static*.

(*b*) It is possible that a system may contain substances or mechanisms such that the forces which do work in a pseudo-static transition are not exclusively those which hold the system in equilibrium. This will be the case if internal frictions are present whose effects do not tend to zero in the pseudo-static limit. It is therefore apposite to single out a special class of pseudo-static transitions, namely *quasi-static transitions*, by the additional requirement that in the course of the latter the work done on K is done precisely by the forces which hold K in equilibrium. It depends on the nature of a given system whether or not it satisfies the condition that every pseudo-static transition is also quasi-static. However, only systems which do satisfy it will henceforth be admitted for consideration.

The requirement that any finite quasi-static transition will require an infinite interval of time must not be taken too seriously. What effectively constitutes a quasi-static transition in practice can ultimately be decided only by an appeal to relaxation times (cf. Section 5). Here, as in any theory involving the consideration of conditions unrealizable in practice, one just has to be 'reasonable' in the interpretation of idealizations which have been introduced: granted that such idealizations do not come into conflict with some basic natural law. The appearance of the Heisenberg Uncertainty Principle in physical theory will serve as a warning against supposing that what may seem 'reasonable' will necessarily be found to be true.

(*c*) If x_k and $x_k + dx_k$ be two states such that

$$|dx_k| < \epsilon \quad (k = 1, 2, \ldots, n), \tag{7.3}$$

where ϵ is some sufficiently small preassigned positive number, then a direct quasi-static transition between them is 'infinitesimal' (note the end of Section 11). In the course of such a transition the mechanical work dW done by K on its surroundings will be

$$dW = \sum_{k=1}^{m} P_k dx_k, \tag{7.4}$$

where m is the number of deformation coordinates. The coefficients P_k are the (*generalized*) *forces* of K; and these may be determined by experiment. (In the simple case considered in Section 5 one has $dW = P_1 dx_1$, where $P_1 = x_2$ if x_2 is the non-deformation coordinate, or $P_1 = x_1^{-2} x_3$ if x_3 is the non-deformation coordinate; all of which amounts to writing $dW = PdV$, in a more familiar notation.) In every case the forces are functions solely of the coordinates $x_1, x_2, \ldots, x_n$, since, according to the definition of quasi-static transitions, they depend on the 'instantaneous' state alone.

In a finite quasi-static transition the external work done by K is

$$W = \int \Sigma P_k dx_k = \int_{t_1}^{t_2} \sum_{k=1}^{m} P_k(f_1(t), \ldots, f_n(t)) \dot{f}_k(t)\, dt, \tag{7.5}$$

recalling (7.1).

(*d*) Commonly one introduces also the notion of *reversible* transitions. It is not unusual to find the terms 'reversible' and 'quasi-static' treated as synonymous. These exists no universal convention in this respect. However, it seems desirable—not least from the viewpoint of some of the later sections of this book—to draw a distinction between them. Accordingly, given a system K, let it first be agreed to refer to everything which is not K as the 'surroundings' of K (see also Section 9*d*). Then a transition from some state $\mathfrak{S}_1$ to some state $\mathfrak{S}_2$ is called reversible if there exists a subsequent transition from $\mathfrak{S}_2$ to $\mathfrak{S}_1$ such that the condition of the surroundings of K is in all respects the same after the second transition as it was before the first. When a transition is not reversible it is called *irreversible*.

Note that nothing has been said explicitly about the character of either transition, e.g. whether it is quasi-static or not. True, it is not difficult to show (see Section 31) that quasi-static transitions are in fact reversible. It is, however, by no means obvious that, conversely, all reversible transitions of thermodynamic systems must be quasi-static; and any statement to the effect that this must be the case would appear to be no more than a conjecture, however strongly one may feel it to be valid. The behaviour of mechanical systems is of course no counter-example.

8. Adiabatic enclosures and partitions

Let a system K in equilibrium be contained entirely within an enclosure which need not be rigid but shall be supposed to be impermeable to matter. Then it is known from experience that it may be possible to disturb the equilibrium of K only by mechanical means. When this is the case the enclosure is called *adiabatic*.

In general K may incorporate a stirrer, or an electrical resistance through which a current may be passed from outside the enclosure, and so on. Then the passage of such a current, movement of the stirrer, variations of the deformation coordinates, are all to be counted as mechanical processes. These, however, belong to only one type of possible interaction between a system and its surroundings. Explicitly, by bringing K into contact with certain other bodies its equilibrium will usually be affected. In other words, a mere change of the condition of the surroundings of K will bring about observable changes within K, distance forces such as electromagnetic forces being excluded for the moment. The enclosure is then not adiabatic, and it is called *diathermic*.

There is, as usual, an element of idealization in these definitions, in as far as it is impossible to realize a strictly adiabatic enclosure in practice. As in the discussion of Section 4 one may well rest content with the pragmatic observation that enclosures can be constructed which approximate an ideal adiabatic enclosure to a high degree. The familiar Dewar flask is just of this kind. If mechanical interference with a system contained in such a flask be excluded, the state of this system cannot be affected at all: a lighted bunsen burner outside the flask brings about no effects within it. Naturally, in saying this one knows very well that it is not quite true in practice; that it represents a good approximation to the actual state of affairs would however be hard to deny, and no issue of principle appears to be involved.

It is useless to define adiabatic enclosures in terms of 'impermeability to heat', for heat has neither been defined, nor is it accessible to direct perception. On the contrary, the definition which has been adopted does not involve any logical circularity; and at the same time it corresponds closely to one's intuitive view as to what distinguishes the two kinds of enclosures under consideration.

One calls a system which is contained within an adiabatic enclosure *adiabatically isolated*. This notion must be kept distinct from that of 'isolation', for the latter implies the absence of *all* interactions with the surroundings, whereas that of mere adiabatic isolation certainly does not exclude those which are mechanical. In short, there are two types of possible interactions between a system and its surroundings, namely 'mechanical' and 'thermal' (meaning: non-mechanical) interactions; and adiabatic isolation implies the absence of the latter. In parenthesis it should be added that electromagnetic and gravitational interactions are to be regarded as mechanical.

The idea of the adiabatic isolation of one system from another rather than from its surroundings now requires scarcely any comment. Two systems are or are not adiabatically isolated from one another according as any interaction between them must be purely mechanical or not. If the mutual adiabatic isolation is physically provided by some material dividing wall then the latter is called an *adiabatic partition*. When convenient, two systems may of course be regarded as two parts of a composite system.

Finally, a given system K may or may not be adiabatically isolated at different times. When it is adiabatically isolated it will be represented by the symbol K_0; and any transition of K_0 is called an *adiabatic transition*. Since the condition of adiabatic isolation does not represent the imposition of a geometrical constraint upon K, the deformation coordinates of K_0 must still be adjustable at will.

9. Concerning the character of thermodynamic theory

(*a*) An appropriate stage has now been reached for a short digression concerning certain features of thermodynamic theory. To begin with, granted the general idea of a thermodynamic system, a moment's contemplation of previous sections reveals a basic emphasis on the notion of equilibrium. Indeed, all the quantitative statements of classical thermodynamics are about states of systems, never about systems not in equilibrium. Consequently any kind of detailed description can be given only of pseudo-static transitions. It is the pre-eminence of the place which static conditions occupy in thermodynamics which has frequently led to the suggestion that the theory should properly be called 'thermostatics'. Despite some virtues which the proposal possesses, it seems to be basically misleading in one important respect. A fundamental ingredient of the theory is the recognition of the existence in nature of irreversible transitions. The notion of an *order in time* is implicit in this; that is to say, under certain definite conditions one can assert which of two states $\mathfrak{S}_1$ and $\mathfrak{S}_2$ of a given system has preceded the other *in time* if only it is known that at some time its state was $\mathfrak{S}_1$ and at some other time $\mathfrak{S}_2$. To the extent that this before-and-after relation is obscured by the term 'thermostatic' its use does not appear as attractive as it might otherwise be. In any case, it is not fruitful to engage in verbal hair-splitting, and the traditional appellation 'thermodynamics' will be retained here.

(*b*) Thermodynamical theory affords an incomplete description of the behaviour of systems, in that it deals with states—widely separated in time—in which they might find themselves, without, in general, being able to give any account of the processes responsible for the transitions between these states, or of the rate at which transitions proceed, as they actually occur in nature. A direct concomitant of this incompleteness is the need for the kind of pragmatic approach recommended, to say the least, in Section 7*b* for instance. Indeed, it is probably true that the very mention there of relaxation times implies a background of a better, more complete theory. This much may be granted; but the purpose of this work is the exposition of a well-established ('classical') theory, not the presentation of a new one.

(*c*) It is usual to distinguish between two different types of thermodynamic theory: 'phenomenological' and 'statistical'. Both concern themselves with the description of, or relationships between, equilibria of systems in the large. The distinction between them is essentially this: phenomenological thermodynamics (which alone is here to be considered) presumes the constituents of any system to be given in bulk, their properties being characterized by a finite, usually small, number of physical quantities, some of which will be included amongst the coordinates of the system. These properties are thus to be accepted as they appear on a macroscopic level, and no attempt is made to relate them to the behaviour of the atomic constituents of the substances which make up the system. As far as phenomenological theory is concerned the question whether the constitution of matter is continuous or discrete does not arise. Consequently, quantum mechanical considerations never appear explicitly, and limitations such as those imposed upon measurability by the Heisenberg Principle are irrelevant. Moreover, there is no place within the strictly phenomenological theory for the idea of fluctuations about equilibrium (see however Section 76).

In statistical thermodynamics the situation is otherwise. Although this discipline also concerns itself with the states (there called 'macro-states') of systems, it does so by contemplating in the first place the so-called 'micro-states' in which the system might find itself. Here a micro-state is, in the classical theory, essentially the entire set of values of the physical quantities, such as the generalized coordinates and momenta, required to describe the dynamical history of the individual atomic constituents of which the system supposedly consists. On a quantum-theoretical level this notion of a micro-state is of course meaningless, and the state-vector (or wave-function) of the system, regarded as an assembly of its elementary constituents, takes its place. In both cases it is recognized that in general a large number of possible micro-states is compatible with the much more scant information provided by any macro-state, and attention is therefore focused on the calculation of suitable averages over micro-states.

It is evident that the input of the statistical theory is far greater than that of the phenomenological theory. Consequently one may expect its output also to contain a greater wealth of detail. In particular, the values of all manner of phenomenological quantities now become predictable, instead of having to be merely taken for granted as revealed by direct observation. Yet this predictability is based on the acceptance of special hypothetical or idealized models: and the extent to which these are fallible is reflected in a corresponding degree of fallibility of the detailed results of the statistical theory.

However this may be, one is dealing with two distinct theories. The conceptual framework of the one differs widely from that of the other, and the mathematical foundations of the statistical theory seem to be a good deal more harassing than those of the phenomenological theory. The fact that the same name may be given to two different quantities of which one appears in the one theory and one in the other does not imply the lack of any distinction between them (cf. also the considerations of Section 1*c*). For example, one quantity arises in the context of adiabatic inaccessibility (cf. Section 38), another is the equilibrium limit of a statistical average whose particular time-dependence conveys an overall feature of the history of a system not in equilibrium. Both of these are called entropy; yet the statistical quantity is given this name because it enters into certain theoretical relationships in just the way in which the corresponding phenomenological quantity enters into certain analogous relationships of the phenomenological theory. To this extent the latter is therefore presupposed. Were it otherwise, the familiar demand that 'statistical thermodynamics must yield results which are not in conflict with the laws of phenomenological thermodynamics' would be meaningless. For the same reason, if no other, the claim that the concepts of the latter theory are 'intelligible' only in terms of those of the former also does not make much sense, though no doubt it is motivated by individual personal demands of 'visualizability'.

(*d*) Finally, a word of warning against the tendency to drag 'the universe' into thermodynamic theory when to do so is neither required nor justifiable. Two distinct aspects of this are invoked in the following remarks. First, one finds the term 'surroundings' in the context in which it occurs in Section 7*d* replaced by the term 'rest of the universe'. True, the surroundings of K were there taken to be as everything that was not K; and it might be argued that one would define the 'rest of the universe' in just the same way. None the less, what is at stake is that only the condition of certain bodies reasonably close to K should be the same at two different times. It does not matter that there is some vagueness inherent in the notion of 'reasonable closeness'. In any special case one can be more precise, and if K is some system within a laboratory for instance, the surroundings will surely be delimited to a terrestrial scale. What goes on elsewhere is irrelevant, the more so as all information concerning the state of distant matter is retarded.

This observation naturally brings the second aspect of references to the universe to mind. Here the applicability of the laws of thermodynamics to large parts of the universe is in question. Now, for laws to be meaningful the concepts inherent in them must be meaningful. It is, however, not immediately clear how the classical concept of energy for instance survives on a cosmical level. Granted that the general relativity theory must be invoked, how is the impossibility of covariantly defining energy to be accommodated? How are deformation coordinates to be defined, especially if the universe as a whole is to be considered? What are the conceptual consequences of the logical impossibility of 'looking at' the universe from outside itself? Many more questions of this kind can be asked, but even this small selection should suffice to suggest that a facile application of laws, framed on the basis of essentially local phenomena, to systems of cosmical extent is not to be undertaken lightly.

10. Standard systems

(*a*) The kind of systems to be admitted for consideration has already been described in preceding sections. Now, as a matter of convenience, it was laid down at the end of Section 3 that the total amount of matter constituting any particular system was to be kept fixed. (Contrary situations will be allowed for later by appropriate generalizations.) Again as a matter of convenience, bearing the aim of simplicity constantly in mind, a further restriction will be placed for the time being on the kind of systems to be contemplated, which relates to the character of its coordinates. Explicitly, of the n coordinates as defined in Section 3, exactly $n-1$ shall be deformation coordinates. The nth coordinate must then of course be a non-deformation coordinate. As a consequence of this restriction such a system cannot have any internal partitions which adiabatically isolate parts of it from each other. That this is so is more easily demonstrated later on, and it will be done in Section 14.

Again, there is no reason in principle why electromagnetic and gravitational forces, or the effects of surface tension, could not be admitted from the outset. However, under some circumstances their inclusion is apt to lead to expositions suffering from a somewhat bewildering appearance; and their generality is not matched by transparency. To the extent to which no issues of principles are at stake, the kinds of interactions just referred to will therefore be disregarded, at any rate for the time being. When all is said and done, the main objective to be attained is the examination of the general consequences of the existence of two different kinds of interactions between a system and its surroundings, as described earlier. In other words, the crucial point is the necessity for the introduction of non-deformation coordinates; together with the concomitant observation that even in an adiabatic transition the work done by K_0 is not uniquely determined by the values of the deformation coordinates of the terminal states. That this is so becomes obvious upon considering the special case of *isometric* transitions, i.e. transitions in the course of which the deformation coordinates have constant values. The simple example of a system K_0 whose interaction with the surroundings is provided by a stirrer will suffice. Work is certainly done in a (non-static) transition of this kind; but the deformation coordinates do not vary at all.

(*b*) Let the kind of system which is being contemplated now be called a *standard system*, and unless otherwise indicated the symbol K will henceforth represent such a system. In short, the essential differences between it and more general systems are contained in the following assumptions:

(i) of the n coordinates just one must be a non-deformation coordinate;

(ii) no substances are present whose properties depend upon their previous histories;

(iii) the effects of surface tension may be neglected;

(iv) mutual long-range interactions between different parts of the system may be neglected;

(v) transitions which occur at an infinitesimal rate are quasi-static, so that, if x_n is the single non-deformation coordinate, the work W done by the system in such a transition is

$$W = \int \sum_{k=1}^{n-1} P_k dx_k, \qquad (10.1)$$

where the forces P_k are functions of $x_1, ..., x_n$ alone. Except under special circumstances a definite path of integration must, of course, be specified.

It may be noted that (iii) and (iv) mainly serve to avoid complications which might otherwise arise when two systems are combined into one composite system. In principle there is always a mutual gravitational interaction between different systems, or between different parts of one system. This is surely negligible in experiments on the laboratory scale, but not so in the context of systems on an astronomical scale. With regard to (ii), one thus excludes, for instance, the presence of water as a working substance (at least for the time being), though one does not normally regard it as falling in this category. However, the point here is that is it known from experiment that two samples of water, both of which have the same density and are at the same pressure may yet under certain conditions exhibit different properties, for example with respect to their refractive indices (see also Section 19*b*). In this regard water is somewhat exceptional.

(*c*) It is sometimes useful to have a straightforward example of a standard system which has n coordinates, $n > 2$. For this purpose it is only necessary to compound such a system out of $n-1$ enclosures in mutual diathermic contact, each of which is entirely filled by a simple fluid. The volumes of these enclosures shall be freely adjustable, and they may be taken as the $n-1$ deformation coordinates $x_1, ..., x_{n-1}$. Any one of the n pressures within the enclosures may be selected to play the role of the non-deformation coordinate x_n. (Of course, even in this case one does not have to choose the $x_1, ..., x_{n-1}$ in just this way. If the kth volume, say, is determined by the position of some lever, then the number which specifies this position may equally well be chosen to be x_k.)

11. Representative spaces

(*a*) The language of thermodynamic theory is greatly simplified by introducing the notion of a *representative space*. (The reader will probably have met these already in the form of '$P-V$ diagrams' or 'indicator diagrams'). The language of ordinary geometry then becomes available as an aid to one's imagination.

Let $x_1, ..., x_n$ be a set of n (real) variables. A set of values of these is called a *point*, the variables themselves *coordinates*. The collection of points corresponding to all possible values of the coordinates within certain ranges constitute an n-dimensional space A_n. (This meaning of the word 'space' has nothing to do with physics.) A *curve* in A_n is defined as the collection of all points generated by the equation

$$x_k = f_k(u) \quad (k=1,\ldots,n), \tag{11.1}$$

where the f_k are continuous functions of the single parameter u. More generally the collection of points

$$x_k = f_k(u_1, u_2, \ldots, u_s) \quad (k=1,\ldots,n), \tag{11.2}$$

with $s < n$, is an s-dimensional *subspace* of A_n, the f_k being continuous functions of the s parameters $u_1, \ldots, u_s$. In particular, when $s = n-1$ the subspace is called a *hypersurface* of A_n. When $n = 3$, $s = 2$ one speaks simply of a surface.

The A_n becomes a metric space if one associates with every pair of points a *distance* between them. This distance will be denoted by $d(x', x'')$, x standing for the entire set of coordinates $x_1, \ldots, x_n$, and x', x'' for their sets of values corresponding to the points in question. The properties one assigns to the 'metric' $d(x', x'')$, regarded as a function of x' and x'', are to some extent a matter of choice. In view of the purposes to which the distance is to be put here it is convenient to lay down that (i) $d(x', x'')$ vanishes if and only if x' and x'' coincide, i.e. $x' = x''$, (ii) $d(x', x'') = d(x'', x')$, and (iii) $d(x', x'') + d(x'', x''') \geqslant d(x', x''')$. These properties of course do not determine a specific function d explicitly. However, one may choose

$$d(x', x'') = \left\{ \sum_{k=1}^{n} (x''_k - x'_k)^2 \right\}^{\frac{1}{2}}, \tag{11.3}$$

for the stated conditions are satisfied by it. With (11.3), and granted that the ranges of the x_k are $-\infty < x_k < \infty$ $(k=1,2,\ldots,n)$, the space is called *euclidean*. It sufficies for later purposes to restrict oneself to such euclidean spaces.

(*b*) It will now be taken for granted that a one-to-one correspondence can be established between the points of a connected domain D of an n-dimensional euclidean space and the possible states in which a given system K whose coordinates are $x_1, \ldots, x_n$ can find itself. Put otherwise, one identifies every state $\mathfrak{S}$ of K

with a point of the (abstract) space, the point $(x_1, \ldots, x_n)$ of the space being identified with the state $\mathfrak{S}$ whose coordinates are $x_1, \ldots, x_n$. When this has been done the space is called a *representative space*, denoted by R_n.

It is now possible to speak indiscriminately of 'states of K' or of 'points of R_n', whichever is convenient, the former being represented by the latter. Alternatively one may say that the representative point of $\mathfrak{S}$ is the 'image' of $\mathfrak{S}$, and this image may also be denoted by the symbol $\mathfrak{S}$. Recalling the definitions of Section 7, and in particular (7.1), one sees at once that the image of a quasi-static transition is a curve $\mathfrak{C}$ in R_n. Further, recall the remark of Section 6 that 'in the course of a non-static transition K is in no state at all'. In terms of R_n this means that the terminal representative points of such a transition are not connected by a representative curve.

It may be noted that reference was made above to a 'connected domain D' of R_n. The first point here is that the set $\{\mathfrak{S}\}$ of all possible states of K may not cover the whole of R_n. A reasonable, or at any rate convenient, choice of coordinates of K will usually be such that one or more of these can only vary over finite ranges. If, for instance, a coordinate is a volume it can certainly not be negative. With regard to the connectedness of D, this means that it shall be possible to join any two of its points by a continuous curve; that is, there shall exist the possibility of some quasi-static transitions between the states they represent. (Of course, no conditions of isolation of K are envisaged at present.)

The metric (11.3) defines a 'distance between states'. It will of course be understood that this has a purely formal significance. Its value depends upon the character of the coordinates which happen to have been chosen, and any change of the units implicit in the assignment of particular values to the coordinates will affect it. However, its utility reveals itself not least in the possibility of attaching a definite meaning to the notion of a 'neighbourhood of a state $\mathfrak{S}$'. First, write $\mathfrak{S}(x)$ for the state in which the coordinates have the values x (i.e. $x_1, \ldots, x_n$), so that the distance between the states $\mathfrak{S}(x)$ and $\mathfrak{S}'(x')$ will be $d(x, x')$. Then, precisely, an (open) ϵ-neighbourhood of a state $\mathfrak{S}(x)$ is the set of all states $\mathfrak{S}'(x')$ such that

$$d(x, x') < \epsilon \quad (\epsilon > 0). \tag{11.4}$$

For the sake of simplicity the phrase 'in every neighbourhood of $\mathfrak{S}$' shall be understood to mean 'in every ϵ-neighbourhood of $\mathfrak{S}$'. For example, the statement that in every neighbourhood of $\mathfrak{S}$ there exists another point $\mathfrak{S}'$ with a certain property means that one can find a point $\mathfrak{S}'$ with this property such that $d(x, x') < \epsilon$, no matter how small ϵ may be.

'Neighbouring states' are such that the distance between them is less than some sufficiently small preassigned positive number ϵ; so that the infinitesimal transition contemplated at the beginning of Section 7*c* is between neighbouring states. A little care must, however, be exercised here, in as far as a transition between neighbouring states need not be infinitesimal, since its representative curve could pass through points of a large region of R_n. In short, if a transition is to be infinitesimal its representative curve shall be a straight line in R_n joining neighbouring representative points.

12. The notations of mathematics and physics

(*a*) In applications of thermodynamic theory the formalism commonly used is noticeably encumbered by the appearance of subscripts attached to partial derivatives. For example, a derivative might appear as $(\partial F/\partial x)_{y,z}$, whereas in investigations of a mathematical nature the provision of round brackets and subscripts is comparatively rare. These formal differences are characteristic of the existence of two distinct notations, namely that of mathematics and that of physics.

As regards the former, contemplate a variable A which is a given function of the variables $x_1, \ldots, x_n$, say,

$$A = F(x_1, \ldots, x_n). \tag{12.1}$$

Its partial derivative with respect to x_k is written $\partial F/\partial x_k$. Let $q(x_1, \ldots, x_n)$ be a second prescribed function of the same variables. Thus, granted single-valuedness, there is associated with every set of values of $x_1, \ldots, x_n$ a number y,

$$y = q(x_1, \ldots, x_n). \tag{12.2}$$

It will be supposed that (12.2) can be resolved uniquely for any one of the $x_1, \ldots, x_n$. Thus,

$$x_k = \phi_k(x_1, \ldots, x_{k-1}, y, x_{k+1}, \ldots, x_n). \tag{12.3}$$

In general the *functions* $\phi_1, \phi_2, \ldots$ are of course quite different from each other. Clearly, if one uses (12.3) to eliminate x_k from F in (12.1) one obtains a new function, i.e. of the variables occurring on the right-hand side of (12.3), and a new functional symbol must be used:

$$A = F(x_1, \ldots, x_{k-1}, \phi_k, x_{k+1}, \ldots, x_n)$$
$$= f_k(x_1, \ldots, x_{k-1}, y, x_{k+1}, \ldots, x_n). \qquad (12.4)$$

Note that for every choice of the particular variable which has been eliminated one has a distinct functional symbol, and so n of them in all.

As regards partial derivatives of A, the strict terminology must include a statement specifying the particular variables on which A is taken to depend, e.g. 'the derivative with respect to x_r of A, regarded as a function of $x_1, \ldots, x_{k-1}, y, x_{k+1}, \ldots, x_n$ $(r \neq k)$'. This derivative, then, is simply $\partial f_k / \partial x_r$, and there is no ambiguity in the notation since the symbol f_k already implies which particular set of variables is being used.

The notation employed in some branches of physics is less explicit as a consequence of the double use of symbols, serving now both as functional symbols and as symbols denoting values of the functions in question. Thus, (12.1) will now be written as

$$A = A(x_1, \ldots, x_n), \qquad (12.5)$$

and (12.2) likewise

$$y = y(x_1, \ldots, x_n). \qquad (12.6)$$

If A be now regarded as a function of $x_1, \ldots, x_{k-1}, y, x_{k+1}, \ldots, x_n$ as in (12.4), one again writes

$$A = A(x_1, \ldots, x_{k-1}, y, x_{k+1}, \ldots, x_n). \qquad (12.7)$$

From the point of view of the notation of mathematics this is strictly speaking nonsense, for (except in the very special case in which $y(x_1, \ldots, x_n)$ happens to be the function x_k) the functional dependence of A on the two sets of variables in question is quite different. Still, all this is merely a matter of convention, and the ambiguity or lack of definiteness apparent from (12.5) and (12.7) is of little account in some contexts, and is counter-balanced by a substantial and convenient economy of symbols. For example the equality of two variables A and B is expressible as $A = B$, irrespectively of the sets of variables on which A and B may have been chosen to depend.

When one now turns to partial derivatives the situation is not quite so happy. For example, the derivative with respect to x_r of A, regarded as a function of the variables occurring in (12.7), clearly cannot be written as $\partial A/\partial x_r$, since these variables are not explicitly indicated. That is to say, in the notation of mathematics $\partial A/\partial x_r$ might mean any one of the derivatives $\partial F/\partial x_r$ or $\partial f_k/\partial x_r$ $(k \neq r)$. Three alternatives for dealing with this difficulty present themselves: (i) any equation involving derivatives is accompanied by a statement explicitly enumerating the variables upon which the various functions depend; (ii) a derivative of the kind $\partial A/\partial x_r$, where A is intended to be regarded as a function of $x_1, \ldots, x_{k-1}, y, x_{k+1}, \ldots, x_n$, say, is written as $\partial A(x_1, \ldots, x_{k-1}, y, x_{k+1}, \ldots, x_n)/\partial x_r$; (iii) the typical derivative just considered is written as $(\partial A/\partial x_r)_C$, where the subscript C stands for the set of variables which 'is kept constant' in the differentiation, i.e. in this case C stands for y and the set $x_1, \ldots, x_n$ other than x_k and x_r. This is the notation most commonly employed in thermodynamic theory, as mentioned at the beginning of this section. Clearly it is entirely equivalent to (ii), in that C and x_r together make up the set of arguments on which, in (ii), A is explicitly stated to depend. Which of the three alternative conventions one uses is of course of no account, though (i) leads to a more attractive appearance of many equations than the other two.

(*b*) A brief remark concerning notation in the context of linear differential forms may not be out of place. A linear differential form in n variables is a linear homogeneous function of n differentials, i.e.

$$dL = \sum_{r=1}^{n} X_r(x_1, \ldots, x_n)\, dx_r. \tag{12.8}$$

Here dL is to be understood as a composite symbol, that is to say, except under very special circumstances no meaning is to be attached to d and L taken disjointly, in the sense of d representing a differential operator which acts on a function $L(x_1, \ldots, x_n)$. In this general situation one sometimes meets with a special notation, the most common of which is represented by the use of $đL$ in place of dL. It will not be used here.

13. Outline of following chapters

The general superstructure of thermodynamics may be erected upon the groundwork of definitions and conceptual tools provided in preceding sections. Accordingly, in Chapters 2–7 of this exposition the four principal laws of the theory, and the characteristic physical quantities to which they give rise, are dealt with in turn. In the very briefest outline, they are concerned with: (i) the condition governing the mutual equilibrium of several systems in diathermic contact, whence the concept of empirical temperature derives; (ii) the limitation imposed upon the amount of external work which can be done by an adiabatically isolated system, and the comparison of the amount of work done by a system in adiabatic and non-adiabatic transitions between given states, leading to the definition of energy and heat in turn; (iii) the phenomenon of adiabatic inaccessibility, which gives rise to the concepts of entropy and absolute temperature; (iv) the behaviour of the entropy function as the absolute temperature tends to zero.

It may be noted that the four laws are dealt with in their traditional order, despite the fact that there are didactic reasons for introducing them in a different order, in particular the Second Law before the two which traditionally precede it. Moreover, the entropy in particular will be defined by diverse means, none of which is, however, based explicitly upon the Kelvin or the Clausius formulations of the Second Law.

The presentation of this basic material is followed in Chapter 8 by the definitions of thermodynamic potentials and constitutive coordinates, together with considerations of a general kind concerning the equilibria of physico-chemical systems. Chapter 9 disposes of a selection of miscellaneous matters of interest. Finally, Chapters 10 and 11 concern themselves with more specialized consequences of the general conclusions reached earlier. These conventional representative applications are severely circumscribed, and are intended to achieve little more than a systematic classification of different types of problems or results within the framework of classical thermodynamics.

CHAPTER 2

THE ZEROTH LAW

14. The mutual equilibrium of two systems

(*a*) Of any two separated (standard) systems K_A and K_B each may be regarded as part of the surroundings of the other. Imagine them to be adiabatically isolated, so that they are in certain states $\mathfrak{S}_A$, $\mathfrak{S}_B$. K_A and K_B may be brought into physical contact, a compound system K_C so being formed. The deformation coordinates of K_A and K_B are to be kept fixed in this process. The adiabatic isolation of these subsystems then implies, according to Section 8, that the states $\mathfrak{S}_A$ and $\mathfrak{S}_B$ will maintain themselves side by side indefinitely. Suppose now that the adiabatic isolation of K_A and K_B be broken just to the extent that these systems are now in diathermic contact, K_C as a whole continuing to be adiabatically isolated. Then, in general, transitions of K_A and K_B will occur, so that when K_C has attained equilibrium their final states $\mathfrak{S}'_A$ and $\mathfrak{S}'_B$ will differ from their initial states. In other words, two arbitrarily prescribed states of K_A and K_B cannot in general coexist when these systems can interact thermally with each other.

In the situation just described the deformation coordinates $x_1, \ldots, x_{n-1}$ and $y_1, \ldots, y_{m-1}$ of K_A and K_B respectively were kept fixed so that in the transitions referred to the non-deformation coordinates x_n and y_m alone took different values in the terminal states. (Of course, $\mathfrak{S}_A$ and $\mathfrak{S}_B$ may just happen to be such that these states can coexist.) In any event, experience shows that quite generally *two systems in diathermic contact will be in equilibrium if and only if their states satisfy one condition of the form*

$$f(x_1, \ldots, x_n; y_1, \ldots, y_m) = 0. \tag{14.1}$$

(*b*) The compound system K_C is again a standard system. To see this it is enough to observe that its state may be taken to be a set of values of the $n+m-1$ coordinates $x_1, \ldots, x_n; y_1 \ldots, y_{m-1}$. The mth coordinate, here chosen to be y_m, is redundant, since its value in any state of K_C is, through (14.1), already implied by the values of the remaining coordinates. It is of course essential that

(14.1) be satisfied. Were the component parts of K_C adiabatically separated, then (14.1) would not be available and none of the coordinates could be eliminated. This means that the states of K_C would involve the values of two non-deformation coordinates, and K_C would not be a standard system. Since a system containing an internal adiabatic partition may be thought of as made up of two subsystems separated adiabatically from each other, it follows that a standard system cannot contain such a partition, as was already stated in Section 10*a*.

The kind of complication that arises when a system K_C^* is not standard on account of its being constituted of, say, two adiabatically separated subsystems K_A and K_B may here be noted. The coordinates shall be $x_1, ..., x_n, y_1, ..., y_m$, with both x_n and y_m non-deformation coordinates. Then it is obvious that if a standard system K with coordinates $z_1, ..., z_l$ be 'brought into diathermic contact with K_C^*' then the value of the non-deformation coordinate z_l, when equilibrium has set in, will depend on what part of K_C^* has been brought into contact with K. All deformation coordinates having been kept fixed the equilibrium value of z_l cannot depend on $x_1, ..., x_n$ if K is in contact with that part of K_C^* to which the coordinates $y_1, ..., y_m$ refer; for the presence of the adiabatic partition which is supposed to exist within K_C^* then entails that under the present conditions there is no interaction of any kind between K_A and K. The mutual equilibrium, therefore, of K and K_C^* cannot be subject to merely a single condition of the kind (14.1). (See also Section 19*c*.)

(*c*) It might be objected at this stage that K_C^* should, except under special circumstances, not be regarded as being in equilibrium at all. If one adopts this viewpoint, namely, that (when K_A and K_B are each separately in equilibrium) K_C^* is only in 'mechanical' but not in 'thermal' equilibrium, some of the definitions given previously require modification. However, it may be reflected that the substitution of a diathermic for an adiabatic internal partition constitutes an essential *modification* of a given system. Yet one would hardly wish to contemplate other kinds of radical modification, such as the removal of some partition. This would again entail a change in the number of coordinates, though in this case the number of deformation coordinates is at stake. It seems better to take the view that as the result of any such modification one is simply left with a new system altogether; for which reason the distinction between certain generically different *conditions* of a system (i.e. whether it be in thermal or mere mechanical equilibrium) is here replaced by the distinction between the possible states of certain generically different *systems* (i.e. standard or non-standard). At any rate, as remarked earlier, attention will be mainly focused on standard systems, in harmony with the usual procedure.

15. The Zeroth Law

In addition to the two systems K_A, K_B of the previous section let a third system K_C be introduced, its coordinates being $z_1,\ldots,z_l$, say. The mutual equilibrium of K_A and K_B is characterized by (14.1), i.e.

$$f(x_1,\ldots,x_n;\, y_1,\ldots,y_m) = 0. \tag{15.1}$$

In the same way the mutual equilibrium of K_B and K_C is subject to a functional relation of the kind

$$g(y_1,\ldots,y_m;\, z_1,\ldots,z_l) = 0, \tag{15.2}$$

whilst corresponding to the mutual equilibrium of K_C and K_A there exists a relation of the form

$$h(z_1,\ldots,z_l;\, x_1,\ldots,x_n) = 0. \tag{15.3}$$

Experiment now teaches that there exists a general relation between the states of three (standard) systems, in the sense expressed by the

> *Zeroth Law: if each of two systems is in equilibrium with a third system then they are in equilibrium with each other.*
>
> (15.4)

This is to be understood as follows. K_A is in an arbitrary fixed state. The states of K_B and K_C are such that upon each of them being brought in turn into diathermic contact with K_A, equilibrium between K_A and K_B on the one hand, and between K_A and K_C on the other, already exists. Then, if K_B and K_C are subsequently brought into diathermic contact with each other, they will necessarily be already in mutual equilibrium. It is of course to be taken for granted that the mere process of moving the various systems about does not sensibly disturb their states. In terms of the equations (15.1–3) the transitivity of the equilibrium relation expressed by the Zeroth Law requires that any two of them imply the third. From this one infers that they must be equivalent to a set of equations of the form

$$\xi(x_1,...,x_n) = \eta(y_1...,y_m) = \zeta(z_1,...,z_l). \qquad (15.5)$$

To see this, observe that if y_r occurs in (15.1) then it must also occur in (15.2), or else this coordinate could not be eliminated from (15.1, 2) taken jointly, to give a relation involving x and z alone. (x, y, z stand, as before, for the entire sets of coordinates of K_A, K_B, K_C respectively). Upon eliminating y_r all other coordinates of K_B must however also be absent from the resulting relation, so that the y_k ($k = 1,...,n$) must occur in (15.1) and (15.2) in the same algebraic combination:

$$f(x, y) = f^*(x, \eta(y)), \qquad (15.6)$$

$$g(y, z) = g^*(\eta(y), z). \qquad (15.7)$$

Whereas f was a function of $n+m$ arguments, f^* is a function of only $n+1$ arguments, i.e. $x_1,...,x_n, \eta$; and an analogous remark applies to g^*. If one now solves the equations $f^* = g^* = 0$ for η one gets just the two equations (15.5). Within certain ranges of the coordinates the function occurring in these are assumed to be continuous and differentiable.

16. The empirical temperature

(*a*) It has been seen that, as a consequence of the Zeroth Law, it is possible to associate with two systems $K_A(x)$ and $K_B(y)$ functions $\xi(x)$ and $\eta(y)$ respectively, such that the condition of mutual equilibrium takes the form

$$\xi(x) = \eta(y). \qquad (16.1)$$

In other words, with every system K, coordinates x, one can associate a function $\xi(x)$, called its *empirical temperature function*, such that two systems, in states $\mathfrak{S}_A$ and $\mathfrak{S}_B$ respectively, are in mutual equilibrium if and only if the empirical temperature functions take the same values for the two states in question. The value of the empirical temperature function of a system K in a state $\mathfrak{S}$ is called the *empirical temperature* of K in the state $\mathfrak{S}$ (or else simply the empirical temperature of $\mathfrak{S}$); and it will be denoted by t. The condition for the mutual equilibrium of two systems K_A and K_B may then be written

$$t_A = t_B. \tag{16.2}$$

It is obvious that the empirical temperature function of a system is not uniquely defined, for the condition (16.1) may equivalently be written in the form

$$\Omega(\xi(x)) = \Omega(\eta(y)), \tag{16.3}$$

where Ω is selected to be any monotonically increasing or decreasing function of one argument. (Requirements of continuity or differentiability on Ω are understood.) Then one might equally well take the empirical temperature of K in the state $\mathfrak{S}$ to be the value of $\Omega(\xi(x))$. This arbitrariness reflects the existence in practice of various 'temperature scales' (see also Section 19*a*).

17. Isothermals

(*a*) The equation

$$\xi(x) = t, \tag{17.1}$$

where t is regarded as a variable parameter, defines a set of hyper-surfaces in R_n. These hyper-surfaces are known as the *isothermals* of K. (If $n = 2$ one has the familiar set of 'isothermal curves'). These isothermals are intrinsically unaffected by the arbitrariness inherent in the definition of $\xi(x)$, to which reference has already been made. As far as the isothermals are concerned, a different choice of Ω corresponds merely to a different parametrization, that is to say, a different way of numbering the hyper-surfaces. The required monotonic property of Ω ensures that their ordering is not disturbed.

(*b*) A word might be said here about the subjective notions of 'hotness' and 'coldness'. It is obvious that the argument developed so far does not suggest any connection between them and temperature. Indeed, it should be reflected that if some particular temperature function is such that, given two states $\mathfrak{S}'$ and $\mathfrak{S}''$, the temperature of the first exceeds that of the second, then by a suitable choice of Ω this relationship can be reversed. No doubt one can so arrange matters that the empirical temperature of water is less than that of liquid iron (both being at atmospheric pressure): but this is merely a question of prescription, not of necessity. Sometimes the notions of hotness and coldness—the hotter body having the higher temperature—are related to questions of energy transfer (see Section 26). However, this way of delimiting acceptable empirical temperature functions has nothing directly to do with the subjective connotations of the terms under discussion. This surely is as it should be. For whether a body is directly felt to be hot or cold depends not only upon its state, but is governed by all manner of subjective factors. In short, it is merely an additional result of experience that if a temperature function is such that it ascribes a higher temperature to liquid iron than to water, then *usually* that one of two bodies subjectively appears the hotter which has the higher temperature.

18. Temperature as coordinate. Equation of state

(*a*) Let a particular temperature function of a system K be given, that is, a definite choice of the arbitrary function Ω shall have been made. (It may be noted that all references to 'temperature' are here of course to empirical temperature, since for quite some time yet no other kind of temperature will have been defined.) Then, in virtue of (17.1), any one of the original coordinates x_k of K may be replaced by t. However, if the resulting set of quantities is to be in harmony with the definition of coordinates of a standard system one will have to eliminate just the non-deformation codinate in favour of t. That having been done, the discussion of compound systems for instance is considerably simplified.

The coordinates $x_1, \ldots, x_{n-1}, t$ may still be denoted collectively by x, the choice $x_n = t$ then being understood. At times it may be appropriate to be more explicit. In that case denote the set of deformation coordinates by $\bar{x}$ $(= x_1, \ldots, x_{n-1})$, so that a state of K may be represented either by $\mathfrak{S}(x)$ or $\mathfrak{S}(\bar{x}, t)$: and a transition which is isometric is characterized by

$$\bar{x} = \text{const.} \tag{18.1}$$

(*b*) It has already been assumed that (17.1) is soluble for any one of the coordinates. Any relation of the form

$$\phi(x_1, \ldots, x_n, t) = 0 \tag{18.2}$$

equivalent to (17.1) is called an *equation of state* of K. On closer reflection it may seem a little strange that a special terminology should attach to a relation of the kind (18.2). After all, any equation

$$\Phi(x_1, \ldots, x_n, u) = 0, \tag{18.3}$$

chosen more or less at random defines u as a function of $x_1, \ldots, x_n$; and the family of 'iso-u hyper-surfaces' generated in R_n by (18.3), when u is treated as a variable parameter, corresponds just to the isothermals (18.2). So (18.3) should be another kind of 'equation of state'. Actually, this is precisely the point of view which is sometimes adopted, except in so far as the function in (18.3) is not chosen quite arbitrarily. Certain specific functions naturally present themselves, for instance that which obtains when u is the energy U of K, defined in Section 22. In that case one does indeed refer to (18.3) also as an equation of state, more exactly a 'caloric equation of state'. Equation (18.2) must of course then be qualified in some way, and one speaks of a 'thermal' equation of state.

The prominence of (18.2) arises from the fact that equations of state are directly concerned with the mutual equilibrium or lack of mutual equilibrium of standard systems. Indeed, the one-to-one correspondence between the isothermals in $R_m(y)$ of a system K_B on the one hand and isothermals in $R_l(z)$ of a system K_C on the other is sometimes employed as a basis for the introduction of temperature. There one considers first the sets of all states of K_B and K_C, separately in equilibrium with a third system K_A which is in a fixed state $\mathfrak{S}_A$; and then a sequence of such fixed states of $\mathfrak{S}_A$ is contemplated. Although this procedure is essentially equivalent to that above, it is perhaps a little closer to the experimental situation.

19. Thermometers

(*a*) At the beginning of the previous section it was supposed that a definite choice of the function Ω had been made. How, in principle, will this be done in practice? Various possibilities are available, one of which may serve as a suitable example. A convenient system K^* is selected, the intention being to bring it into diathermic contact with various other systems in turn. K^* is then limited to isometric transitions, that is to say, it is agreed to keep all its deformation coordinates $\bar{x}^*$ fixed. Recalling (17.1) and (16.3), t is then a function of the single coordinate x_n^*, say

$$t = \Omega(f^*(x_n^*)). \tag{19.1}$$

Consequently one may simply take x_n^* itself to be the empirical temperature t, and this choice corresponds exactly to taking Ω as the function inverse to f^*. One may equally well take any monotonic function of x_n^* in place of x_n^* itself: this is merely a matter of convenience. At any rate, having once decided upon any such particular function, the empirical temperatures of any other systems are determinable, and no arbitrariness remains. Thus, if x_n^* itself has been chosen as the empirical temperature, the temperature of any other system in diathermic contact with K^* is the value of x_n^*; and if the temperatures of two systems have been found to be equal in this way, then these systems will be in mutual equilibrium when brought into contact with each other, in virtue of (15.5). All this is of course closely related to the remark made at the end of Section 18, K_A there playing the role of K^*.

In the example above the states of K^* were restricted to differ only in the values of the non-deformation coordinate; that is to say, all states admitted for consideration were required to lie on a line in R_n. More generally, one selects a suitable K^* and imposes such conditions upon it as will ensure that the image of the set of its possible states is a curve in R_n, as before. Then K^* is called a *thermometer*.

In practice one sometimes chooses K^* to be a suitable gas contained in an enclosure of fixed volume, x_n^* being the pressure of the gas. This constitutes the so-called 'constant-volume gas thermometer'. An even more familiar example perhaps is the mercury-in-glass thermometer, albeit a rather complex one. Therefore, rather than start from first principles, it is simpler to observe that there exists a one-to-one correspondence between the constant-volume gas thermometric temperature and the length of the mercury column; granted that the glass enclosure is rigid in the sense that the application of external mechanical stresses will not sensibly affect its shape. Once this is established the gas thermometer may simply be discarded.

The length of the mercury column, or any monotonic function

of it, will then serve as empirical temperature or, as one also says, a particular 'temperature scale'. Of course, one will not normally wish to consider 'strange' scales, as exemplified by the choice $t = e^{-kl}$, for instance, where k is a positive constant and l the length of the column. On the contrary, one will usually proceed with the considerations of Section 17*b* in mind. Yet it must not be thought that the choice of $t = kl$ is necessarily a convenient one; it certainly would not be if the capillary tube had a highly non-uniform bore. At any rate, further remarks concerning temperature scales will fit more appropriately into the context of Section 49.

(*b*) A thermometer must be a standard system, and in Section 10*b* water was excluded as a working substance. One can now see more easily why this step was necessary: in general one does not have a one-to-one correspondence between the constant-volume gas temperature t_g defined above and the length of the water column in a water-in-glass 'thermometer'. Such an instrument therefore is not really a thermometer at all. One has, indeed, the situation that two samples of water, the pressures and densities of which are equal, may yet have different temperatures t_g. (At normal pressure the density of water attains a maximum value at about 4 °C.) Of course, this phenomenon occurs only for a certain range of t_g, outside of which water may be admitted as part of a standard system. However, one may now go further and admit it even within this range provided one agrees not to recognize a pair of values of pressure and volume as being a state of it. On the contrary, a state of water shall include its temperature. In short, water need no longer be excluded from consideration; but the existence of two distinct families of isothermals which overlap in a part of its representative space preclude its use for the purpose of *defining* an empirical temperature after the fashion of Section 16.

(*c*) The remarks of Section 14*b* concerning a non-standard system K_C^* which consists of two standard subsystems which are adiabatically isolated from each other may be slightly enlarged upon with the observation that one evidently cannot ascribe a single empirical temperature to K_C^*. Rather, two temperatures have to be associated with it if the possible equilibria of K_C^* with any other system are to be described after the manner of preceding sections. If K_C^* has r internal adiabatic partitions one will likewise require $r+1$ temperatures.

One should be strictly conscious of the fact that *temperature is defined only under conditions of equilibrium.* No temperature can be assigned to a system undergoing a non-static transition. If one attempts to define the temperature of a system not in equilibrium by some prescribed procedure involving a thermometer, the results arrived at would not be unique in as far as they depend upon irrelevant and hardly controllable features of the measuring instrument. Thus the reading of a mercury-in-glass thermometer which is exposed to sunlight will depend on whether its bulb happens to be painted black or white. Incidentally, if an isothermal transition is defined by the condition that t shall be constant throughout its course then the notion of a non-static isothermal transition is meaningless.

There are differential equations which describe the time dependence of the 'temperature' at arbitrary points of a given body K^*. Clearly the equation is concerned with non-equilibrium conditions, and it would not appear to be legitimate to speak of temperature at all. This is true. Nevertheless, it is a matter of experience that results obtained under widely different circumstances by means of a diversity of measuring instruments are coherent to within experimental error. Despite the absence of equilibrium a local temperature may therefore be assigned in terms of the readings of a suitable thermometer. The operative part of this must be sufficiently small now, but otherwise it is to be used as if one were dealing with conditions of equilibrium. Of course, if the local readings are not constant in time ('non-stationary' conditions) even this does not seem very meaningful. However, whether the conditions be stationary or not, one can help oneself with the following conceptual artifice. At any chosen time τ the system is subdivided 'instantaneously' into an infinite (i.e. sufficiently large) number of adiabatically isolated subsystems. If the point a falls within the subsystem K_a^*, then the temperature at a at time τ is the temperature of K_a^* at a time τ' sufficiently later than τ for equilibrium to have set in. Evidently thermal conduction falls outside the range of description of classical thermodynamics proper. Nevertheless, it will be considered in a little more detail in Section 75.

(*d*) A final remark concerns the assumed continuity and single-valuedness of the various functions which were introduced in this chapter. Whatever coordinates enter into the description of a system K, the continuity ensures that for any positive number η there exists an ϵ-neighbourhood of any state $\mathfrak{S}$ all states $\mathfrak{S}'$ of which satisfy $|t'-t| < \eta$. Consequently, if initially one used a representative space R_n whose points are images of states of K (coordinates $\bar{x}$, x_n) then one may equally well employ in place of R_n another space R_n' to represent states of K when the coordinates are taken as $\bar{x}$, t. Of course $d(x, x')$ will in general differ numerically from

$$d'(x, x') = \left\{(t-t')^2 + \sum_{k=1}^{n=1} (x_k - x_k')^2\right\}^{\frac{1}{2}}, \qquad (19.2)$$

but this is of no concern: R_n' in fact satisfies whatever general conditions were imposed upon R_n.

CHAPTER 3

THE FIRST LAW

20. Laws of conservation

Observed regularities in the behaviour of things are reflected in the formulation of physical laws. Given a number of these it may turn out that they reveal certain common or regular features. By way of a simple example, the basic equations of a number of different field theories might involve only derivatives of at most the second order. Such an observed rule is sometimes elevated to the status of a 'law'. Since it is not concerned directly with the behaviour of things, but with formal aspects of the modes of description of these, one has rather a law about physical laws; and, for the sake of clarity at least, this may be called a 'metalaw'. It might be noted that a clear example is furnished by the Restricted Principle of Relativity, when it is stated (somewhat loosely) in the form 'laws of physics shall be invariant under Lorentz transformations'.

A prominent place is occupied in many physical theories by laws of conservation. These express directly permanent features of the world, in that conservation of a physical quantity Q is synonymous with its constancy in time: though the system to which the quantity relates may be undergoing complicated changes, the value of Q will under appropriate conditions not be affected by these. Thus in classical mechanics the components of the total linear momentum of an isolated set of particles are conserved, i.e. they have the same values before and after these particles have undergone mutual collisions. Likewise, the total energy of this system is conserved, provided the system is purely mechanical, i.e. the collisions are elastic. The importance of such conserved quantities lies in their availability even when the detailed history of the system may be unknown.

As remarked earlier, when a system is not isolated the work done by external forces reappears either as energy of motion or else is represented by the occurrence of elastic deformations. When the kinetic energy and the elastic deformations disappear the work which originally produced them is recovered. Energy is therefore conserved provided one ascribes to elastically deformed bodies a potential energy, equal to the work done in producing their deformations. This is to be understood in the slightly more general sense that the total energy, i.e. the sum of the kinetic and potential energies of a mechanical system, changes in any interactions with its surroundings by an amount just equal to the work done on it. In particular, it is constant when the system as a whole is isolated.

When there is no (tangible) mechanical action on a system its energy may appear to be not conserved. This is so, for instance, when a rigid body is moving in a (Newtonian) gravitational field. In that case it is possible to ascribe to it another kind of potential energy which depends upon its position in the field, in such a way that, once again, its total energy will be conserved. However, when time-dependent fields come in to play this is no longer good enough, for the values of the field at the places where a particle is located may be zero at two different times, yet the values of the kinetic energy of the particle at these times may differ from each other. To rescue the principle that the energy of mechanical systems is conserved one then has to look upon the field as constituting a mechanical system in its own right, and an energy density is ascribed to the region of space in which the field is not zero.

The crucial issue at stake here is that one has had to define new physical quantities, regarded as 'forms of energy', in order to be able to maintain a conservation law earlier formulated under more restrictive circumstances. Evidently, the motivation is provided by the acceptance of a metalaw: according to which one can associate with a general mechanical system a conserved quantity called its energy E, such that any external work done on the system is just balanced by the increase of E. This is a metalaw, as one is concerned with a limitation upon the laws governing the histories of mechanical systems; these laws having to be such that it shall be possible to define E. It is of course by no means obvious *a priori* that their structure will be of the required kind. In classical electrodynamics one is already confronted with certain difficulties in this context. More strikingly, one may have adopted other metalaws, the consequences of which are in conflict with the metalaw under discussion; and the Principle of General Covariance may here serve as an example.

According to the remarks of Section 2 it is just the existence of non-conservative systems which motivates thermodynamic theory in the first instance. Yet the availability of a conservation law for energy is such a desirable feature of a theory—hence the metalaw just discussed— that the question arises immediately as to whether one might not after all be able to ascribe to a thermodynamic system an energy which is at least conserved in a *restricted class* of transitions. The following two sections devote themselves to this problem; but reference may also be made to Sections 28 and 29.

21. The First Law

Amongst possible restricted classes of transitions which one might hope to be relevant to the search for a conservation law as motivated in the preceding section, that of adiabatic transitions immediately springs to mind. One need only reflect that the adiabatic isolation of a system K_0 in one respect restores its purely mechanical character, namely just to the extent that (by definition) all interactions between it and its surroundings can only be mechanical. It is therefore natural to study the work W_0 done by K_0 in a transition between arbitrarily prescribed terminal states $\mathfrak{S}'(x')$, $\mathfrak{S}''(x'')$. Such transitions may proceed in any manner whatever, that is to say, they certainly need not be quasi-static; and in practice they may be brought about by the stirring of fluids as well as by mere adjustments of the deformation coordinates at arbitrary rates. The results of many such experiments then find their expression in the

> *First Law: the work done by a system K_0 in an adiabatic transition depends on its terminal states alone.* (21.1)

Let it be emphasized again that (i) this law makes reference only to *adiabatic* transitions, and that (ii) the work done by K_0 is independent of the manner in which the transition proceeds. That W_0 depends only on $\mathfrak{S}'$ and $\mathfrak{S}''$ means that it is a function of x' and x'' only. One may therefore write

$$W_0 = F(x'; x''). \tag{21.2}$$

For the present it does not matter if the system has internal adiabatic partitions.

22. The energy

(*a*) Let $\mathfrak{S}'(x')$, $\mathfrak{S}''(x'')$, $\mathfrak{S}'''(x''')$ be any three states, ordered in such a way that the transitions of K_0 from $\mathfrak{S}'$ to $\mathfrak{S}''$, and from $\mathfrak{S}''$ to $\mathfrak{S}'''$, and therefore from $\mathfrak{S}'$ to $\mathfrak{S}'''$, shall be possible. Then $\mathfrak{S}''$ may be referred to as an 'intermediate state'. The work done in the transition from $\mathfrak{S}'$ to $\mathfrak{S}'''$ is $F(x'; x''')$; and this is the sum of the amounts of work done in first bringing the system from $\mathfrak{S}'$ to $\mathfrak{S}''$ and then from $\mathfrak{S}''$ to $\mathfrak{S}'''$. Thus, identically for any intermediate state $\mathfrak{S}''(x'')$,

$$F(x'; x'')+F(x''; x''') = F(x'; x'''). \tag{22.1}$$

In forming the sum on the left the variables x'' must therefore cancel out, and this requires that $F(x', x'')$ be the difference between a function of x' alone and the *same* function of x'' alone:

$$F(x'; x'') = U(x')-U(x''). \tag{22.2}$$

The First Law thus directly associates with any thermodynamic system K a certain physical quantity U which is a function only of the coordinates of K, such that the difference between the values of U, calculated for any two states, is equal to the work done by K in any adiabatic transition between these states. This quantity U is known as the *internal energy function* of K, or more simply 'the energy' of K. One may at times speak of the energy of the state $\mathfrak{S}$, meaning the value of the energy function of K when the latter is in the state $\mathfrak{S}$; but no confusion is likely to arise when one neglects to use this more precise terminology.

It is convenient to write the right-hand member of (22.2) alternatively also as $U'-U''$ or $-\Delta U$; it being understood that if X is any function of the coordinates of K, then

$$\Delta X = X(x'')-X(x') \tag{22.3}$$

is the difference between the values of X associated with the final and initial states of a transition respectively. Equation (22.2) then reads

$$W_0+\Delta U = 0, \tag{22.4}$$

and this is just the formal expression of the sought-for conservation law, valid under the anticipated condition (see also Sections

28 and 29*a*). It should be noted that U is defined only to within an arbitrary additive constant. This may be fixed by assigning some convenient value to some chosen state $\mathfrak{S}^*$. Then the energies of all other states are uniquely fixed relative to $\mathfrak{S}^*$.

(*b*) If two states $\mathfrak{S}'$ and $\mathfrak{S}''$ of K_0 be arbitrarily prescribed it may happen that no transition from $\mathfrak{S}'$ to $\mathfrak{S}''$ is possible. Indeed, as will be discussed at length later on, this situation is of common occurrence. The question therefore arises as to whether a transition from $\mathfrak{S}''$ to $\mathfrak{S}'$ then necessarily exists. In general the answer to this is in the negative. To see this one need only imagine an exact replica of K_0 to have been constructed, which together with K_0 forms a compound system K_C. This will be non-standard on account of the presence of an internal adiabatic partition. Now consider a state $\mathfrak{S}'_C$ of K_C in which K_0 is in the state $\mathfrak{S}'$ and its replica in the state $\mathfrak{S}''$, and alternatively a state $\mathfrak{S}''_C$ in which K_0 is in $\mathfrak{S}''$ but its replica in $\mathfrak{S}'$. Then $\mathfrak{S}'_C$ and $\mathfrak{S}''_C$ are obviously inaccessible from one another. Under these circumstances any assertion about 'the work done in a transition between these states' would be meaningless, and $U''_C - U'_C = \Delta U$ has a less direct interpretation. In fact U'_C and U''_C are separately defined in terms of other states from which $\mathfrak{S}'_C$ and $\mathfrak{S}''_C$ can be reached, or which can be reached from them. ΔU is then a mere 'paper and pencil quantity'.

An analogous situation might conceivably exist in the case of standard systems. However, in harmony with experience, one makes the assumption that

If $\mathfrak{S}'$, $\mathfrak{S}''$ are arbitrarily prescribed states of an adiabatically enclosed standard system such that transitions from $\mathfrak{S}'$ to $\mathfrak{S}''$ are impossible then there exist transitions from $\mathfrak{S}''$ to $\mathfrak{S}'$. (22.5)

Equation (22.5) has the status of an *ancillary law*. Its validity is often taken for granted without comment, whilst sometimes it is claimed to be deducible with the aid of certain consequences of the Second Law. However, the arguments which are adduced for this purpose appear either to involve some circularity, or else to require the validity of additional assumptions. Thus, one accepts that

Given any state $\mathfrak{S}(\bar{x}, U)$ of an adiabatically enclosed standard system, every state of the continuous sequence $\mathfrak{S}'(\bar{x}, U')$, $U' > U$, can be reached from $\mathfrak{S}$. (22.6)

This reflects the possibility of increasing the energy of K_0 through the action of frictional forces; but in general it is only necessary that $\mathfrak{S}'$ should be isometric with $\mathfrak{S}$, i.e. the transition itself need not be isometric. At any rate, it may be noticed that (22.5) is a consequence of (22.6). Thus, let $\mathfrak{S}'(\bar{x}', U')$, $\mathfrak{S}''(\bar{x}'', U'')$ be the states to which (22.5) refers. If $\mathfrak{S}'$ be taken as the initial state, let the deformation coordinates be adjusted quasi-statically to their desired final values $\bar{x}''$. Whatever the value of the energy U^* of the state $\mathfrak{S}^*(\bar{x}'', U^*)$ which has so been reached may be, one can assert that it must be greater than U'' for otherwise $\mathfrak{S}''$ could be reached from $\mathfrak{S}^*$ (and therefore from $\mathfrak{S}'$), in view of (22.6). If, therefore, one now takes $\mathfrak{S}''$ as the initial state one may first go isometrically to the state $\mathfrak{S}^*$. Anticipating here the result of Section 31 that quasi-static adiabatic transitions are reversible one can then go from $\mathfrak{S}^*$ to $\mathfrak{S}'$ by reversing the transition considered at the outset. This shows that (22.6) implies (22.5). Though (22.6) has not been required hitherto, its validity will henceforth be accepted.

23. Heat

Consider again arbitrarily prescribed states $\mathfrak{S}'$, $\mathfrak{S}''$. So long as K is adiabatically isolated any transition between $\mathfrak{S}'$ and $\mathfrak{S}''$ is characterized by equation (22.4). Let the condition of adiabatic isolation now be removed, and contemplate *any* transition between these same states $\mathfrak{S}'$, $\mathfrak{S}''$. ΔU, of course, is already known, whereas the work W done can be measured. $W+\Delta U$ will now usually fail to vanish and it is convenient to introduce the special symbol Q for this quantity, by way of abbreviation:

$$Q = \Delta U + W. \tag{23.1}$$

Whatever the value of Q may happen to be for any particular transition, one traditionally refers to it as the *heat* absorbed by K in this transition. Despite the quaint terminology one should of course not think of some 'substance' being 'absorbed'. Q is *defined* by (23.1); in words: *the heat absorbed by K in any particular transition between given states is the difference between the work actually done by K in this transition and the work which it would have done had the transition between the given terminal states been adiabatic.*

When the First Law is taken for granted, but not until then, adiabatic transitions may be characterized by the absence of transfer of heat to or from K.

24. Additivity of the energy

Let K_A, K_B be two adiabatically enclosed standard systems, and let $\mathfrak{S}'_A$, $\mathfrak{S}'_B$ and $\mathfrak{S}''_A$, $\mathfrak{S}''_B$ be two pairs of states such that $\mathfrak{S}''_A$ can be reached from $\mathfrak{S}'_A$, and $\mathfrak{S}''_B$ from $\mathfrak{S}'_B$. Further, these various states shall be restricted by the imposition of the conditions $t'_A = t'_B$ and $t''_A = t''_B$ on their temperatures. An adiabatically enclosed compound system K_C may be formed by bringing K_A and K_B into contact with each other. $\mathfrak{S}'_A$ and $\mathfrak{S}'_B$ jointly define an initial state $\mathfrak{S}'_C$ of K_C, whilst $\mathfrak{S}''_A$ and $\mathfrak{S}''_B$ likewise define a final state $\mathfrak{S}''_C$ of K_C. The work W_C done by K_C in the transition from $\mathfrak{S}'_C$ to $\mathfrak{S}''_C$ is the sum of the amounts of work done by K_A and K_B, and these in turn are $-\Delta U_A$ and $-\Delta U_B$ respectively. The truth of this conclusion is ensured by the assumptions made earlier that distance and surface forces are absent or at least negligible, so that the energy functions of K_A and K_B are the same after mutual contact has been established as they were whilst they were separated. On the other hand, in view of the definition of energy, $W_C = -\Delta U_C$, so that

$$\Delta U_C = \Delta U_A + \Delta U_B.$$

With a proper choice of the additive constant left free in U_C it therefore follows that

$$U_C = U_A + U_B. \tag{24.1}$$

So far K_C has not been a standard system on account of the adiabatic partition between K_A and K_B. If this be replaced by one which is diathermic, either initially or finally, the states of K_A, K_B and K_C are in no way affected. According to the First Law (21.1) the work W_C done by K_C in an adiabatic transition between given terminal states is independent of the manner in which the transition is carried out. Now, let it be granted that the phrase 'the manner in which the transition is carried out' is meant to include the possibility of temporarily modifying a given system by the insertion of an adiabatic partition. It then follows that if K_A and K_B were in diathermic contact not only initially and finally but throughout the transition from $\mathfrak{S}'_C$ to $\mathfrak{S}''_C$ the work done on K_C

was just equal to the amount W_C above. Selecting $\mathfrak{S}'_C$ to be some suitable standard state the result (24.1) now relates the energy of a compound system to the energies of its constituent systems quite generally ('additivity of energy'), subject to the previous assumptions concerning surface and distance forces. In a general situation it then follows from (24.1) that

$$Q_C = Q_A + Q_B, \tag{24.2}$$

since Q, like U, may be defined separately for K_A and K_B, and for K_C as a whole.

25. On certain properties of the energy function

(*a*) In this chapter the coordinates of a system have not hitherto been required to fulfil any conditions other than those implicit in the definition of standard (sub-) systems. It may therefore be noted incidentally that the statement of the First Law in no way presupposes the notion of temperature, so that it might quite reasonably precede the statement of the Zeroth Law. However this may be, an empirical temperature t being available, it will be convenient to take the non-deformation coordinate of K to be in fact just t, as suggested in Section 18*a*. This entails no loss of generality in view of the remarks of Section 19*d*.

One assumes the energy function $U(x)$ to be continuous and differentiable—at least for certain ranges of the coordinates. As regards continuity, one is in effect making a physically plausible assumption concerning the amounts of work done by a system in adiabatic transitions between neighbouring states. Thus, let the work done by K_0 in the transition from a state $\mathfrak{S}$ to a state $\mathfrak{S}'$ be W. If $\mathfrak{S}''$ is any state in an ϵ-neighbourhood of $\mathfrak{S}'$, and the work done in a transition from $\mathfrak{S}$ to $\mathfrak{S}''$ is $W+\delta$, then it is assumed that

$$\delta \to 0 \quad \text{as} \quad \epsilon \to 0. \tag{25.1}$$

(*b*) That the energy is a *single-valued* function of the coordinates is a conclusion already contained in the statement of the First Law; so that the isoenergetic hyper-surfaces U = const. in R_n cannot intersect. (Recall that implicit in this is the admissibility and sufficiency of the coordinates, and therefore in turn the absence of any substance whose properties depend on its previous history.)

If a particular coordinate x_k $(k \neq n)$ is singled out, all others being taken as constant, x_k need however not be a single-valued function of U. The situation is different with regard to t, in consequence of (22.6). The sequence of states $\mathfrak{S}'(\bar{x}, U')$ to which it refers is one of increasing energy. If t' did not vary monotonically with U' one could find two states having the *same* coordinates $\bar{x}$, t, yet such that in the transitions from $\mathfrak{S}$ to these states the amounts of work done on K_0 would differ from one another, in contravention of the First Law. (One cannot argue analogously in the case of the other coordinates since in doing work on K_0 by varying x_k $(k \neq n)$ only, t will not remain constant in general, i.e. one is confronted with simultaneous changes in more than one coordinate.)

It may be noted that (22.6) is not intended to imply that U has no *upper* bound, i.e. that there exists no state of K or of K_0 whose energy exceeds a certain value (the energy of some standard state having been prescribed). However, that no such upper bound exists is a separate assumption which will be adhered to. On the other hand, it is a fact of experience that the energy of every system has a *lower* bound.

26. Energy and empirical temperature scale

In R_n the states of the sequence $\{\mathfrak{S}'\}$ considered in Section 25*b* lie on a straight line $\mathfrak{L}$ through $\mathfrak{S}$, parallel to the t-axis. How any particular $\mathfrak{S}'$ was attained from $\mathfrak{S}$ is irrelevant. However, $\mathfrak{L}$ may be regarded as the image of an isometric pseudo-static (though of course not quasi-static), transition. Recall that for this t is either a monotonically increasing or a monotonically decreasing function of U. As mentioned in Section 17*b* this relationship may be exploited to fix one characteristic of those empirical temperature functions which are conventionally preferred. Thus if $U = U(\bar{x}, t)$ one generally requires t to be such that it *increases* with U:

$$\partial t/\partial U > 0. \tag{26.1}$$

Now this conventional limitation upon the temperature scale has been achieved by considering only a particular sequence $\{\mathfrak{S}'\}$ of a particular system, K_A say; and it is necessary to show that (26.1) will then hold for any arbitrarily selected system K_B. That this is so may easily be inferred as follows. Denote the derivative

$\partial U/\partial t$ generically by c. Let K_A^* be a given system for which $c_A > 0$. K_A shall be geometrically similar to K_A^*, the volume of each of its enclosures being γ times that of the corresponding enclosure of K_A^*, whilst if a certain quantity of some substance is contained in an enclosure of K_A^* then the corresponding enclosure of K_A shall contain γ times the quantity of the same substance. Then c_A is equal to γc_A^*. Suppose now there exists a system K_B for which $c_B < 0$. Then by a suitable choice of γ one can arrange the compound system K_C formed of K_A and K_B to have $c_C = 0$, contrary to the previous conclusion that $|c| > 0$ for any system, so that $c_B < 0$ cannot hold.

By way of amplification it may be remarked that if $c < 0$ were to hold for some system then most strange consequences would arise. For example, let c be respectively positive and negative for K_A and K_B. If t_A is greater than t_B and the systems are brought into diathermic contact equilibrium will establish itself. This process is intended to be isometric, K_C as a whole being adiabatically isolated. Under these conditions

$$W_A = 0, \quad W_B = 0, \quad Q_A + Q_B = 0, \tag{26.2}$$

whence

$$\Delta U_A = -\Delta U_B. \tag{26.3}$$

It follows that with the assumptions made above the temperatures of K_A and K_B would either rise or both fall, so that the equilibrium temperature t'_C of K_C would certainly not lie between t_A and t_B. Moreover, the magnitudes of the members of (26.3) would have to tend to infinity as $|c_A|$ approaches $|c_B|$ more and more nearly; all of which is quite unphysical. Conversely, the fact that c has the same sign for all systems ensures that under the conditions to which (26.2, 3) refer, the final temperature t'_C of K_C is intermediate to the temperatures t_A and t_B; and (since $c > 0$) energy (heat) has in fact been transferred from the 'hotter' to the 'colder' body. With regard to a matter of definition, what has happened is, in effect, this: arguing on the basis of the First Law (together with (22.6)), one has shown that empirical temperature scales may be selected in such a way that states of *any* system of given shape (i.e. $\bar{x} = \text{const.}$) for which the values of the energy form an increasing sequence will have higher and higher temperatures; and one then says that of two states that one is the *hotter* which has the higher temperature. A concomitant of this is then that the diathermic establishment of equilibrium between two systems of given shape proceeds by the transfer of energy (heat) from the initially hotter to the initially colder system.

27. Calorimetric experiments

Conventionally one often begins to discuss the subject of thermodynamics from the point of view of 'calorimetric experiments'. What is such an experiment? In principle—using the language of previous sections—it is the bringing together of several bodies $K_A, K_B, \ldots$, in states $\mathfrak{S}_A, \mathfrak{S}_B, \ldots$ to form a compound standard system K^*, adiabatically isolated as a whole. Each of the bodies will then undergo some transition until K^* attains a state $\mathfrak{S}^*$. Thereafter the empirical problem is to relate $\mathfrak{S}^*$ to $\mathfrak{S}_A, \mathfrak{S}_B, \ldots$. In short, a *calorimeter* is an enclosure which is such as to ensure that the transitions of any system contained within it are both adiabatic and unaccompanied by the performance of work done on the surroundings of the enclosure. Therefore, in any 'calorimetric transition'

$$Q^* = \Delta U^* = W^* = 0. \tag{27.1}$$

Sometimes one envisages the imposition of the individual conditions

$$W_A = W_B = \ldots = 0. \tag{27.2}$$

Then $\Sigma Q_A = 0$, and $Q_A = \Delta U_A$, $Q_B = \Delta U_B, \ldots$, so that the 'heat lost by one system is just equal to the heat gained by all the others'. Here one may just as well replace the term 'heat' by 'energy'; and because of (27.2) there is a one-one correspondence between the amounts of heat lost or gained by any one body and its change of temperature. It is this relationship which formed the subject of early investigations concerning the nature of heat. On the simplest level a solid body is brought into contact with a fluid of some sort within a calorimeter and the resulting changes of temperatures are observed. These are then interpreted under the assumption that the changes of volume which actually occur are so small as not effectively to violate the conditions (27.2).

The remarks of this section may appear rather trivial, as, indeed, they are in the present setting. This is merely a pointer to the redundancy of purely calorimetric experiments in a general thermodynamic setting. To some extent this applies then also to the so-called 'mechanical equivalent of heat'. Suppose the temperatures of a number of systems K_A, K_B, ... are increased isometrically by Δt_A, Δt_B, ... by bringing each in turn into diathermic contact with a certain system K^*, the temperature of the latter decreasing each time from t_1^* to t_2^* ($\bar{x}^* = \text{const.}$). Then the *same* changes $\Delta t_A, \Delta t_B, \ldots$ can be produced by doing the *same* amount of work $-W$ adiabatically on $K_A, K_B, \ldots$ in turn, granted that the initial and final values of the deformation coordinates are to be the same in each case. To make any sense out of the mechanical equivalent of heat one then has to imagine that in the first case some fixed ill-defined 'quantity of heat' 'flowed' from K^* into $K_A, K_B, \ldots$; this heat having *under the present* conditions the substantial property of being conserved, whilst its effects can be produced 'equivalently' by doing the work $-W$. If one chooses some quite definite system K^*, along with a set of values $\bar{x}^*$ of its deformation coordinates and two definite temperatures t_1^*, t_2^*, then one automatically defines a certain amount of work $-W$ (equal to $\mathcal{J}$ say), this being of course merely the increase of energy of K^* in its transition from $\mathfrak{S}^*(\bar{x}^*, t_1^*)$ to $\mathfrak{S}^*(\bar{x}^*, t_2^*)$. $\mathcal{J}$ is then the mechanical equivalent of heat. Strictly speaking this is redundant, and it is sufficient to talk about energies alone.

28. Perpetual motion of the first kind

In works concerning the history of science one sometimes meets accounts of attempts to produce 'perpetual motion machines'. How are such machines to be understood? In the first place there is of course no question that one is not interested in having machines or objects merely carrying out some never-ending motion. After all, two otherwise isolated particles moving in a vacuum under the influence of their mutual gravitational attraction will to all intents and purposes continue to do so indefinitely. No, what is intended is to have *useful* perpetual motion. That is to say, a 'perpetual motion machine of the first kind' would be a system K which interacts with its surroundings $\bar{K}$ only mechanically and which does a positive amount of work W on $\bar{K}$ in the course of a 'closed' transition, i.e. one whose terminal states are identical.

This is, however, impossible since the simultaneous conditions $Q = 0$ and $\Delta U = 0$ are incompatible with $W > 0$, by (23.1). The First Law thus declares immediately the impossibility of constructing perpetual motion machines of the first kind; and the persistent failure of all attempts to construct them is just evidence for the validity of the First Law. It may be remarked that another type of perpetual motion machine makes its appearance in a different context. This, too, proves to be unrealizable, as will be shown in Section 47*a*.

29. Two-sided and one-sided conservation laws. Laws of impotence

(*a*) If $Q = 0$, not only is $W > 0$ incompatible with the requirement that a transition be closed, i.e. that its terminal states be identical, but $W < 0$ is likewise incompatible with it; and each conclusion comes from the equality of the initial and final values of the energy. Given any system K one may, if necessary, always include its surroundings, or at least such parts of it, that any transitions of the composite system K^* are of necessity adiabatic, i.e. $Q^* = 0$. Then for a closed transition $W^* = 0$, and one verbalizes this result by saying that 'energy is neither created nor destroyed'. This conservation law is *two-sided*: both creation *and* destruction of a certain physical quantity are declared to be impossible. However, not all conservation laws need be as strong. On the contrary, one may have *one-sided* conservation laws, which will express the impossibility of either creating or destroying some physical quantity, but not both. An example of this will be met with later: see Section 47*b* in particular.

(*b*) The First Law as formulated originally in Section 21 might indeed be restated either as a two-sided conservation law, or else in the form of a 'Law of Impotence'. A Law of Impotence is a declaration that certain physical processes or operations, though apparently conceivable and meaningful, are in fact impossible. Thus, it does not seem to be *a priori* impossible to construct a perpetual motion machine of the first kind. Going beyond this example, it is a curious and interesting fact that many of the basic laws of physics do indeed have the form of Laws of Impotence. Apart from that above, and another connected with the Second Law of Thermodynamics, one might here cite: (i) the impossibility of distinguishing by means of local experiments one inertial frame of reference from another; (ii) the impossibility of distinguishing by means of local experiments the effects of the acceleration of a local frame of reference from the gravitational effects of matter in its neighbourhood; (iii) the impossibility of assigning exact measured values simultaneously to certain pairs of dynamical quantities. The reader will recognize these as occupying central positions in the Restricted Theory of Relativity, the General Theory of Relativity, and Quantum Mechanics, respectively.

30. Quasi-static adiabatic transitions

(*a*) If a transition is quasi-static (10.1) applies to any infinitesimal part of it. Further, the energy difference of neighbouring states is infinitesimal and one writes dU in place of ΔU, keeping in mind that dU is a total differential:

$$dU = \sum_{k=1}^{n-1} \frac{\partial U}{\partial x_k} dx_k + \frac{\partial U}{\partial t} dt. \tag{30.1}$$

Further, the value of Q for an infinitesimal transition is infinitesimal, and one writes dQ instead. The definition of an infinitesimal transition in Section 11*b* should here be recalled. If a transition is merely between neighbouring states Q certainly need not be small. Recall also the remarks of Section 12*b*: both dQ and dW generically denote merely linear differential forms. Then, from (23.1),

$$dQ = \sum_{k=1}^{n} X_k dx_k, \tag{30.2}$$

where, with $x_n \equiv t$,

$$\left.\begin{aligned} X_k &= P_k + \frac{\partial U}{\partial x_k} \quad (k = 1, \ldots, n-1), \\ X_n &= \frac{\partial U}{\partial t}. \end{aligned}\right\} \tag{30.3}$$

In any finite quasi-static transition

$$Q = \int dQ. \tag{30.4}$$

(*b*) A finite quasi-static transition is called adiabatic if every element of it is adiabatic, i.e. $dQ = 0$, not merely $Q = 0$. Such a transition is therefore characterized by the condition

$$\sum_{k=1}^{n} X_k dx_k = 0; \tag{30.5}$$

and this is a 'linear total differential equation'.

It is apposite to think about this equation in the context of the representative space R_n. Any transition of the kind in question, leading from an initial state $\mathfrak{S}$ to a final state $\mathfrak{S}'$ has as its image a curve $\mathfrak{C}_0$ in R_n, linking the points $\mathfrak{S}$ and $\mathfrak{S}'$. Any infinitesimal part of $\mathfrak{C}_0$ defines an infinitesimal displacement (i.e. vector) $d\mathbf{s}$ the components of which are $dx_1, dx_2, \ldots, dx_n$. At any point of $\mathfrak{C}_0$ the X_k have known values, and $X_1, X_2, \ldots, X_n$ may be thought of as defining a given vector $\mathbf{X}$. The condition (30.5) may therefore be stated as follows: if the element $d\mathbf{s}$ contains a point $\mathfrak{S}''$ of $\mathfrak{C}_0$ then it must be perpendicular to the vector $\mathbf{X}$ at $\mathfrak{S}''$. This requirement of mutual orthogonality of $d\mathbf{s}$ and $\mathbf{X}$ of course does not prevent $\mathfrak{C}_0$ from having 'kinks', i.e. points at which it is not differentiable. However, one naturally requires the number of such kinks to be finite.

Any particular $\mathfrak{C}_0$ may be defined after the fashion of (11.1). The set of functions $f_k(u)$ then constitutes a solution of the differential equation (30.5) in the sense that the $f_k(u)$ are everywhere continuous, and

$$\sum X_k \frac{dx_k}{du} = \sum_{k=1}^{n} X_k(f_1(u), \ldots, f_n(u)) \dot{f}_k(u) = 0 \tag{30.6}$$

wherever they are differentiable. The curve $\mathfrak{C}_0$ is the representative of the solution, and one calls it a *solution curve* of the equation. The general discussion of (30.5) will be taken up fully in the next chapter, so that there is no need to go any further now.

31. Reversibility of quasi-static transitions

The question as to the relationship between the properties (of a transition) of being quasi-static on the one hand and reversible on the other was already referred to in Section 7*d*, and the time has come to take it up again. Of particular importance is the conclusion that every quasi-static adiabatic transition is in fact reversible. Although, in accordance with the remarks of Section 7*d*, it is not *necessary* to show that a transition is reversible in all its details if it is to be reversible as a whole, it is certainly *sufficient* to do so, if this be possible. Let $\mathfrak{C}_0$ be (i.e. represent) a quasi-static adiabatic transition from $\mathfrak{S}$ to $\mathfrak{S}'$. Since K_0 is a standard system its deformation coordinates $\bar{x}$ are, by hypothesis, freely adjustable (see the end of Section 8). Now the transition from $\mathfrak{S}$ to $\mathfrak{S}'$ will have been achieved by varying the deformation coordinates in a certain way (at an infinitesimal rate). In other words the functions $f_1, f_2, \ldots, f_{n-1}$ in (30.6) are prescribed, and the temperature t of any state of $\mathfrak{C}_0$ is therefore provided by the solution of (30.6), regarded as an ordinary first-order differential equation for f_n. The functions $f_1, \ldots, f_{n-1}$ and the initial point $\mathfrak{S}$ thus define the curve $\mathfrak{C}_0$ uniquely; and $\mathfrak{S}'$ must have been given in such a way as to lie on this curve.

If u'' corresponds to some state $\mathfrak{S}''$ of $\mathfrak{C}_0$ then a change du of u induces changes $dx_k = \dot{f}_k(u)\,du$ of x_k $(k \neq n)$, whilst dt is given by

$$X_n\,dt = -\sum_{k=1}^{n-1} X_k\,dx_k, \tag{31.1}$$

where $X_1, \ldots, X_n$ are evaluated at $u = u''$. Moreover, these quantities are unaffected by a mere reversal of sign of all the dx_k $(k \neq n)$ simultaneously. (This would not be true, for instance, for a transition involving frictional forces which do not go to zero in the quasi-static limit since then X_k would depend on the first derivatives of the f_k as well.) It follows from (31.1) that the only effect on dt will be a mere reversal of sign. One concludes that if, having attained $\mathfrak{S}'$, one adjusts the deformation coordinates in such a way that their values run through those of the transition from $\mathfrak{S}$ to $\mathfrak{S}'$ but in the reverse order, then those of t will do likewise. Hence the values of all the coordinates will simultaneously resume those of the state $\mathfrak{S}$. At the same time, since $\Delta U = 0$ and $Q = 0$ together entail $W = 0$ the surroundings of K_0 will also have been restored to their initial conditions, granted of course that no supererogatory changes have occurred there. The reversibility of the type of transition under discussion has thus been demonstrated. (See also Section 37*a*.)

If the system is not adiabatically isolated the question of the reversibility of quasi-static transitions is now somewhat trivial. One first has to grant again that in the interaction of K with its surroundings $\bar{K}$ the latter can be regarded as also changing quasi-statically. Then the change of the composite system consisting of K and $\bar{K}$ will be quasi-static and can certainly be taken as adiabatic. Proceeding in much the same way as above one infers that non-adiabatic quasi-static transitions of K are reversible.

CHAPTER 4

INTEGRABILITY

32. Introductory remarks

The prominence of linear differential forms in classical thermodynamics has already begun to reveal itself in previous chapters, especially in Section 30. It arises ultimately from the fact that the possibility of arriving at quantitative results concerning the details of a transition requires that all coordinates should be defined in the course of it, and then further, that the actual rate at which the system passes through intermediate states shall be irrelevant provided that it is infinitesimal. Just the class of quasi-static transitions must therefore be central to the theory. The possible histories of an adiabatically isolated system in particular are then subject to (30.5), or put otherwise, to the 'equation of motion' (30.6) in which the 'velocities' $\dot{f}_k(u)$ occur *linearly*.

In view of all this it would appear to be desirable to discuss linear differential forms and equations of the type (30.5) in some detail, the more so as one method of developing the consequences of the Second Law (Chapter 5) rests heavily upon the existence of two distinct classes of such forms, as will be explained shortly. The mathematics involved is quite elementary, and acquaintance with it broadens one's understanding of the whole subject. This is true even if one avoids a direct appeal to the characteristic differences between the two classes of forms, when dealing with the general consequences of the Second Law after the manner of Chapter 6. Moreover, much of the detail may, if desired, be omitted at first reading, or altogether, if certain general results are then merely accepted where they are required in Chapter 5; but the reader may wish to avoid the relevant sections of this in any case. The virtue of interpolating purely mathematical detail lies in the clearer separation it affords between the mathematical content of the theory and the physical presuppositions which underlie it.

The following treatment is perhaps a little long-winded. This feature of it may, however, help towards an easier appreciation of its main points; the more so as the theorem of Section 36 is dealt with in a way which it fits more readily than usual into the framework of the discussions of linear differential forms, as commonly found in elementary works on differential equations. (*Remark*: the reader who does not intend to work his way through the whole of Chapters 4 and 5 should read Sections 33, 37 and 38, and then turn directly to Chapter 6.)

33. The two classes of linear differential forms

(*a*) Generically a linear differential form dL is a linear homogeneous function of the differentials of n independent variables $x\ (= x_1, x_2, \ldots, x_n)$:

$$dL = \sum_{i=1}^{n} X_i dx_i, \tag{33.1}$$

the coefficients X_i being functions of $x_1, \ldots, x_n$. The remarks of Section 12*b* should here be recalled, that is to say, dL must in the first place be regarded as a composite symbol, and the notation does *not* imply the necessary existence of a function $L(x)$ such that the expression on the right of (33.1) is its total differential. Before enlarging on this remark it is desirable to introduce a more convenient notation for partial derivatives than that used hitherto. Thus, if R is a function of x_i, its partial derivative $\partial R/\partial x_i$ will be written throughout this chapter as $R_{,i}$. In other words, the partial derivative of a function with respect to x_i is denoted by the functional symbol in question, to which a subscript i preceded by a (subscript) comma has been attached. The prior presence of other subscripts does not affect this notation. By way of example, the derivative of X_i with respect to x_j will be $X_{i,j}$. (Second derivatives do not require an additional comma, e.g. $\partial^2 R/\partial x_i \partial x_j$ is written $R_{,ij}$.) Also, the limits of summation in expressions such as (33.1) will henceforth be left understood, where this is not likely to lead to confusion.

Suppose now that dL happens to be the total differential of a function $R(x)$, i.e.

$$dL = R_{,i} dx_i. \tag{33.2}$$

Comparing (33.2) with (33.1), the arbitrariness of the dx_i implies $X_i = R_{,i}$. However, $R_{,ij} = R_{,ji}$ holds identically, whence

$$X_{i,j} - X_{j,i} = 0 \quad (i, j, = 1, 2, \ldots, n). \tag{33.3}$$

In general these conditions will of course not be satisfied. (Note than when $n = 3$ (33·3) states, in the language of elementary vector calculus, that the curl of a vector is zero if the vector is the gradient of a scalar. From the point of view of geometrical intuition one may do well to think largely in terms of the case $n = 3$.) That dL is not in general a total differential is not at all surprising since there is no reason why n arbitrarily prescribed functions X_i should in every case be derivable by differentiation from a single function. At any rate, (33·3) are the necessary and sufficient conditions for dL to be a total differential. (The sufficiency has not been demonstrated, but this is easily done by means of a simplified version of the method of Section 34.) When

$$X_i = R_{,i}, \tag{33·4}$$

the equation

$$dL = 0 \tag{33·5}$$

reduces to $dR = 0$, which has the solution

$$R = k, \tag{33·6}$$

where k is a constant of integration. This solution is represented in the usual euclidean representative space R_n by a family of hypersurfaces, one of which passes through any given point. Equation (33·6) thus constitutes a (single) *algebraic equivalent* of (33·5), granted that (33·3) is satisfied.

(*b*) Are there any other conditions under which (33·5) has a single algebraic equivalent? The answer to this is affirmative, for (33·6) will still obtain if dL is merely proportional to a total differential instead of being equal to it, i.e. if there exist functions $q(x)$ and $R(x)$ such that

$$dL = q\,dR. \tag{33·7}$$

If this is the case

$$X_i = qR_{,i}, \tag{33·8}$$

whence, by forming the appropriate derivatives, one infers easily that

$$C_{ijk} \equiv X_i(X_{j,k}-X_{k,j})+X_j(X_{k,i}-X_{i,k})+X_k(X_{i,j}-X_{j,i}) = 0. \tag{33·9}$$

Thus, if dL is to have the property in question it is necessary that the C_{ijk} vanish identically; and in Section 34 it will be shown that these 'conditions of integrability' are also sufficient. Evidently only a single condition, viz. $C_{123} = 0$, obtains when $n = 3$, and in elementary vector notation this may be written

$$\mathbf{X} \cdot \operatorname{curl} \mathbf{X} = 0. \tag{33.10}$$

(*c*) When (33.9) is not satisfied (33.5) has no single algebraic equivalent, i.e. no solution of the form (33.6). At one time it was thought that under these circumstances the differential equation (33.5) has no solution and was therefore meaningless. dL is therefore called *non-integrable*, whilst when there is a single algebraic equivalent (i.e. when (33.9) is satisfied) dL is called *integrable*. Of course it is nonsense to say that (in the first case) the equation has no solution, as is already obvious from Section 30*b*. However, the traditional terminology, though somewhat misleading, survives. That is, a linear differential form dL is said to be integrable or non-integrable according as one can or cannot find functions q and R such that, identically, $dL = q\,dR$. Two distinct general classes of forms are thus defined.

(*d*) When q and R exist, q is said to be an *integrating* denominator of dL, since the form resulting from division of dL by q is a total differential. Alternatively, $1/q$ is called an *integrating factor* of dL. Note that q and R are not uniquely defined. For if some pair of such functions q, R has been found, then

$$q^* = q/g'(R), \quad R^* = g(R) \tag{33.11}$$

is an equivalent pair, where g is an arbitrary function of R and g' its derivative with respect to R.

The special case $n = 2$ is somewhat trivial since every linear differential form in two variables is integrable. This is not surprising, for one has two functions q, R to accommodate just two functions X_1, X_2. Of course $dL = 0$ may simply be written as

$$\frac{dx_1}{dx_2} = -\frac{X_2}{X_1}, \tag{33.12}$$

which is an ordinary first-order differential equation.

34. Sufficiency of integrability conditions

The conditions of integrability (33.9) have only been shown to be necessary. For later purposes it is now to be demonstrated that they are also sufficient. Thus, given that the coefficients of the form (33.1) satisfy (33.9) it is to be shown that the form is integrable. This is quite a simple matter, and as a first step the equation $\Sigma X_i dx_i = 0$ will be suitably transformed by the introduction of new independent variables. To this end, regard momentarily all the variables $x_1, \ldots, x_{n-2}$ as constant, so that dL reduces to the sum of two terms $X_{n-1}dx_{n-1} + X_n dx_n$. Since for the present only two 'variables' occur, this is integrable, i.e. functions w and H exist such that

$$X_{n-1}dx_{n-1} + X_n dx_n = w\,dH \quad (x_1 \ldots, x_{n-2} = \text{const.}). \quad (34.1)$$

Of course w and H will in general depend parametrically upon $x_1, \ldots, x_{n-2}$ since X_{n-1} and X_n do so. w and H having been found, let the conditions of the constancy of $x_1, \ldots, x_{n-2}$ be dropped. Then

$$X_{n-1}dx_{n-1} + X_n dx_n = w\left(dH - \sum_{i=1}^{n-2} H_{,i}dx_i\right). \quad (34.2)$$

The equation $dL = 0$ becomes

$$\sum_{i=1}^{n-2} (X_i/w - H_{,i})dx_i + dH = 0. \quad (34.3)$$

Now introduce in place of the variables $x_1, \ldots, x_n$ new variables $y_1, \ldots, y_n$ as follows:

$$y_i = x_i \quad (i = 1, \ldots, n-2, n), \qquad y_{n-1} = H(x_1, \ldots, x_n). \quad (34.4)$$

Then (34.3) takes the generic form

$$\sum_{i=1}^{n-2} Y_i dy_i + dy_{n-1} = 0. \quad (34.5)$$

$Y_1, \ldots, Y_{n-2}$ will in general depend upon *all* the variables $y_1, \ldots, y_n$; but (34.5) is more convenient chiefly on account of the absence of one of the differentials.

Since only a mere algebraic substitution is involved in the passage from the original to the transformed equation, the former is integrable if and only if the latter is. Now, by hypothesis, (33.9) is here satisfied, and so the corresponding conditions on (34.5) are satisfied, i.e.

$$Y_i(Y_{j,k}-Y_{k,j})+Y_j(Y_{k,i}-Y_{i,k})+Y_k(Y_{i,j}-Y_{j,i}) = 0, \quad (34.6)$$

where the differentiations are of course with respect to the new variables. Choose $j = n-1$, $k = n$. Then since $Y_{n-1} = 1$, $Y_n = 0$ (34.6) reduces to

$$Y_{i,n} = 0 \quad (\text{all } i). \quad (34.7)$$

This shows that the Y_i do not depend on y_n, and the form on the left of (34.5) is in fact one in $n-1$ variables only.

The whole procedure may now be repeated as many times as required until one ultimately arrives at a form which involves only two variables. This, however, is certainly integrable. It follows that the original equation, and therefore dL, is integrable if the conditions (33.9) are satisfied; that is, the conditions of integrability are both necessary and sufficient.

35. Character of solutions

(*a*) According to the discussion of Section 30*b* any particular solution of the differential equation

$$\Sigma X_i dx_i = 0 \quad (35.1)$$

is represented by a curve $\mathfrak{C}_0$ in R_n. This conclusion arises directly from the interpretation of (35.1) as an expression of the mutual orthogonality of the vector **X** and the infinitesimal displacement $d\mathbf{s}$; and it involves no considerations of integrability. A curve may be looked upon as the intersection of $n-1$ hyper-surfaces, so that in general the integral equivalent of (35.1) will be a set of $n-1$ algebraic equations:

$$g_k(x_1, \ldots, x_n) = 0 \quad (k = 2, \ldots, n). \quad (35.2)$$

Indeed $n-2$ of these may be chosen arbitrarily, subject to the requirement that $g_k(x_1' \ldots, x_n') = 0$ $(k = 3, \ldots, n)$, where $x_1', \ldots, x_n'$ are the coordinates of an initial point $\mathfrak{S}'$ through which $\mathfrak{C}_0$ is to pass. Then the $n-2$ variables $x_3, \ldots, x_n$ may be expressed as functions of x_1 and x_2,

$$x_j = h_j(x_1, x_2) \quad (j = 3, \ldots, n), \quad (35.3)$$

and they can then be eliminated from (35.1). There results an equation of the generic form (33.12) whose solution is in effect the first member ($k = 2$) of (35.2), the constant of integration having been so chosen that $g_2(x_1', x_2') = 0$. (g_2 having been found in this way it will of course not depend on $x_3, \ldots, x_n$.)

All this is particularly clearly visualizable when $n = 3$. Any solution curve $\mathfrak{C}_0$ of (35.1) passing through $\mathfrak{S}'$ is the intersection of two surfaces containing $\mathfrak{S}'$. Choose one of these arbitrarily; then the other surface is obtained by integrating (35.1), having first eliminated say x_3 from it, as described above.

How are these general conclusions to be reconciled with the fact that when (35.1) is integrable it appears to have an algebraic equivalent consisting of a single equation only, represented by a hyper-surface $\mathfrak{D}$, not a curve? The contradiction is only apparent: the general algebraic equivalent is still (35.2), but when g_2 is determined exactly as above its form will be such that the equation

$$R(x_1, x_2, \ldots, x_n) = R(x_1', x_2', \ldots, x_n') \qquad (35.4)$$

(where $dL = q\,dR$), is implied by (35.2). This must clearly be so, since when $x_3, \ldots, x_n$ are eliminated from $q\,dR$ one after all has the same equation only differently expressed; and its solution must be

$$R(x_1, x_2, h_3(x_1, x_2), \ldots, h_n(x_1, x_2)) = \text{const.} \qquad (35.5)$$

In this case therefore the equation $g_2 = 0$ is simply

$$R(x) = \text{const.} = R(x'). \qquad (35.6)$$

The important conclusion that has been reached is thus the following. Any solution of (35.1) which passes through a prescribed initial point $\mathfrak{S}'$ is represented in R_n by a curve through $\mathfrak{S}'$; but when the equation is integrable any such curve must lie in a *fixed* hyper-surface $\mathfrak{D}$ which contains $\mathfrak{S}'$. The qualification 'fixed' is intended to mean that this hyper-surface is determined by the given coefficients $X_1, \ldots, X_n$ of the equation (and by $\mathfrak{S}'$) alone, that is to say, independently of the choice of the arbitrary functions $g_3, \ldots, g_n$.

For example, suppose $n = 3$, and

$$\begin{aligned} dL &= dx_1 + (x_2/x_1)\,dx_2 + (x_3/x_1)\,dx_3 \\ &= x_1^{-1} d(x_1^2 + x_2^2 + x_3^2). \end{aligned}$$

Then all solution curves of (35.1) which pass through the point 1, 0, 0 are the curves of intersection of the surface

$$g(x_1, x_2, x_3) - g(1, 0, 0) = 0 \quad (g \text{ arbitrary})$$

and the 'fixed' unit sphere $x_1^2 + x_2^2 + x_3^2 - 1 = 0$.

(*b*) Suppose now that $\mathfrak{S}'$ and $\mathfrak{S}''$ are arbitrarily prescribed points and one inquires whether it is possible to find a solution curve of (35.1) which passes through both $\mathfrak{S}'$ and $\mathfrak{S}''$. Then, in view of the general conclusions above, one can immediately assert that if the equation is integrable it will usually *not* be possible. For, given $\mathfrak{S}'$, the equation determines a fixed hyper-surface through $\mathfrak{S}'$ in which all solution curves through $\mathfrak{S}'$ must lie; so that unless $\mathfrak{S}''$ happens to lie within this hyper-surface no curve joining $\mathfrak{S}'$ and $\mathfrak{S}''$ can be found which satisfies the equation. Thus in every neighbourhood of $\mathfrak{S}'$ there are points $\mathfrak{S}''$ which cannot be reached from $\mathfrak{S}'$ along solution curves of (35.1), namely all those points which do not lie in $\mathfrak{O}$.

The integrability of the equation (35.1) has been shown to be a *sufficient* condition for the existence in every neighbourhood of any point $\mathfrak{S}'$ of points $\mathfrak{S}''$ inaccessible from $\mathfrak{S}'$ along solution curves of the equation. The basic question now arises whether the condition is also *necessary*. The affirmative answer to this is contained in a theorem which forms the subject of the next section.

36. The Theorem of Carathéodory

The answer to the question raised at the end of the last section is provided by the

> *Theorem of Carathéodory: If every neighbourhood of any arbitrary point $\mathfrak{S}'$ contains points $\mathfrak{S}''$ inaccessible from $\mathfrak{S}'$ along solution curves of the equation $\Sigma X_i dx_i = 0$ then the equation is integrable.* (36.1)

To prove this theorem it will be shown that given a point $\mathfrak{S}'$ every point in a certain neighbourhood of it is accessible along solution curves of (35.1) unless the integrability conditions (33.9) are satisfied. The presupposition of the existence of inaccessible neighbouring points therefore implies that (33.9) is satisfied, and, in view of the result of Section 34, that the linear differential form dL is integrable.

Let, then, $\mathfrak{S}'(x')$ be some arbitrarily given point, and let $\mathfrak{C}_0$ be a solution curve of (35.1) through $\mathfrak{S}'$. Its equation may be written as in Section 11, i.e.

$$x_i = f_i(u) \quad (i = 1, 2, \dots, n), \tag{36.2}$$

where the f_i are continuous single-valued functions of u. In the present context these may also be taken to be differentiable. Since the set of functions $f_i(u)$ satisfies (35.1), one must have

$$\sum_{i=1}^{n} X_i \dot{f}_i = 0, \tag{36.3}$$

where a dot denotes differentiation with respect to u. Since all summations which occur here are over the same range from 1 to n, summation signs will be omitted from now on in this section, with the understanding that if a subscript i occurs twice in any product summation over the range in question is implied. The value u' of u shall correspond to the point $\mathfrak{S}'$, whilst the value u^* of u shall correspond to some other point $\mathfrak{S}^*$ on $\mathfrak{C}_0$, $|u'-u^*|$ being sufficiently large. One may take u^* to be less than u', and in the range $u^* \leqslant u \leqslant u'$ distinct values of u shall correspond to distinct points.

Now let $\mathfrak{C}_0^*$ be an arbitrary solution curve through $\mathfrak{S}^*$ which lies in the neighbourhood of $\mathfrak{C}_0$, that is to say, a curve whose equation is of the form

$$x_i = f_i(u) + \epsilon\phi_i(u), \tag{36.4}$$

where ϵ is a sufficiently small positive number and the $\phi_i(u)$ are single-valued continuous differentiable functions of u. By hypothesis

$$\phi_i(u^*) = 0, \tag{36.5}$$

and the functions (36.4) satisfy (35.1). Inserting them in this equation and rejecting all terms of degree higher than the first in ϵ one gets the equation

$$X_i \dot{\phi}_i + X_{i,j} \dot{f}_i \phi_j = 0. \tag{36.6}$$

As in the case of the index i repetition of the subscript j implies that summation over it is to be carried out. Of the n functions ϕ_i $n-1$ may be chosen arbitrarily, and then the nth is to be obtained from (36.6). Accordingly let all but ϕ_k be chosen in some way, k being henceforth a *fixed* index, and in all summations over

j the value k is then to be omitted. ϕ_k is now the solution of the ordinary linear first-order equation

$$X_k\dot{\phi}_k + X_{i,k}\dot{f}_i\phi_k = -X_j\dot{\phi}_j - X_{i,j}\dot{f}_i\phi_j. \qquad (36.7)$$

In the usual way one determines an integrating factor of this, i.e. a function $\lambda(u)$ such that after multiplying (36.7) throughout by λ the left-hand member has the form

$$d(\lambda X_k\phi_k)/dt \equiv \lambda X_k\dot{\phi}_k + \lambda X_{k,i}\dot{f}_i\phi_k + \dot{\lambda}X_k\phi_k. \qquad (36.8)$$

λ therefore is the solution of the elementary equation

$$X_k\dot{\lambda} = (X_{i,k} - X_{k,i})\dot{f}_i\lambda. \qquad (36.9)$$

Integrating (36.7) after multiplication by λ, one obtains

$$\lambda X_k\phi_k = -\int_{u*}^{u} \lambda(X_j\dot{\phi}_j + X_{i,j}\dot{f}_i\phi_j)\,du, \qquad (36.10)$$

where (36.5) has been used on the left. The first term on the right may be transformed by means of an integration by parts:

$$\begin{aligned} -\int_{u*}^{u} \lambda X_j\dot{\phi}_j\,du &= -\lambda X_j\phi_j + \int_{u*}^{u} (\dot{\lambda}X_j + X_{j,i}\dot{f}_i)\phi_j\,du \\ &= -\lambda X_j\phi_j + \int_{u*}^{u} \lambda\dot{f}_i\phi_j[(X_{i,k} - X_{k,i})X_j/X_k + X_{j,i}]\,du. \qquad (36.11) \end{aligned}$$

In the first step (36.5) was used again, in the second $\dot{\lambda}$ was eliminated by means of (36.9). Equation (36.10) now becomes

$$\phi_k = -X_j\phi_j/X_k - (\lambda X_k)^{-1}\int_{u*}^{u} (\lambda\dot{f}_i\phi_j/X_k) \times [X_j(X_{k,i} - X_{i,k}) + X_k(X_{i,j} - X_{j,i})]\,du. \qquad (36.12)$$

Because of (36.3) a term $X_i(X_{j,k} - X_{k,j})$ may be added within the square brackets, and then (36.12) finally becomes

$$\phi_k = -X_j\phi_j/X_k - (\lambda X_k)^{-1}\int_{u*}^{u} \lambda X_k^{-1}\dot{f}_i\phi_j C_{ijk}\,du. \qquad (36.13)$$

Now choose $u = u'$, so that the corresponding point $\mathfrak{S}''$ on $\mathfrak{C}_o^*$ can be made to lie arbitrarily close to $\mathfrak{S}'$ by taking ϵ sufficiently small. On the other hand, $\phi_k(u')$ depends not only on the $\phi_j(u')$ $(j \neq k)$, but on the values which the functions $\phi_j(u)$ take in the entire interval $u^* \leqslant u \leqslant u'$. It follows that, unless the integrand in (36.13) vanishes everywhere in this interval, one can choose these *arbitrary* functions ϕ_j $(j \neq k)$ in such a way that $\mathfrak{S}''$ is any point near $\mathfrak{S}'$ for which $d(x', x'') < \eta$, where η is a sufficiently small positive number. Hence all points in an η-neighbourhood of $\mathfrak{S}'$ can be reached from $\mathfrak{S}'$ along solution curves of (35.1), viz. in the manner just described. It follows that the existence of points in this neighbourhood which cannot be so reached requires that

$$C_{ijk}\dot{f}_i = 0. \qquad (36.14)$$

Now of the n function $\dot{f}_i$ all but one may be chosen arbitrarily, and one may take this to be $\dot{f}_k$. However, $\dot{f}_k$ does not occur in (36.14), and so the latter requires

$$C_{ijk} = 0, \qquad (36.15)$$

i.e. the integrability conditions must be satisfied. In view of the result of Section 34, the Theorem of Carathéodory is thus proved.

CHAPTER 5

THE SECOND LAW (I)

37. Adiabatic inaccessibility

(*a*) It was already pointed out in Section 22*b* that if $\mathfrak{S}$, $\mathfrak{S}'$ are arbitrarily prescribed states of an *adiabatically enclosed* system K_0, it may be impossible for K_0 to undergo any transition from $\mathfrak{S}$ to $\mathfrak{S}'$. In the case of a standard system a transition from $\mathfrak{S}'$ to $\mathfrak{S}$ is however possible under these circumstances, according to (22.5). It is of course taken for granted that the impossibility of effecting transitions between the given states in either direction is a peculiarity inherent in the behaviour of the system itself, and not in that of its surroundings. That is to say, any irreversibility of exterior devices operating in the course of transitions of K_0 is to be regarded as irrelevant.

It should be noted that if after the adiabatic transition from $\mathfrak{S}'$ to $\mathfrak{S}$ one restores the original state $\mathfrak{S}'$ by some non-adiabatic process then one will inevitably be left with an overall change in the condition of the surroundings of K, as will be shown in Section 46. In other words, the transition from $\mathfrak{S}'$ to $\mathfrak{S}$ was irreversible, in the general sense of the definition of Section 7*d*.

(*b*) If $\mathfrak{S}$ is any state of K_0 then the existence of states $\mathfrak{S}'$ which cannot be reached from $\mathfrak{S}$ is a general feature of thermodynamic systems; and it will be dealt with at length presently. For this purpose it is convenient (though by no means necessary) to have a special terminology and notation at hand. Thus if a state $\mathfrak{S}'$ can be reached from a state $\mathfrak{S}$ then it is said to be *accessible* from $\mathfrak{S}$; in the contrary case it is *inaccessible*. The statement '$\mathfrak{S}'$ is accessible from $\mathfrak{S}$' will be represented symbolically by $\mathfrak{S} < \mathfrak{S}'$; whilst its negation '$\mathfrak{S}'$ is inaccessible from $\mathfrak{S}$' is written $\mathfrak{S} \nless \mathfrak{S}'$. In the first case the question as to whether $\mathfrak{S}$ is or is not accessible from $\mathfrak{S}'$ is left open. However, if definitely $\mathfrak{S} < \mathfrak{S}'$ *and* $\mathfrak{S}' < \mathfrak{S}$ then one writes $\mathfrak{S} = \mathfrak{S}'$; which means that the transition from $\mathfrak{S}$ to $\mathfrak{S}'$ is reversible. [It should be constantly borne in mind that the symbolic statement '$\mathfrak{S} = \mathfrak{S}'$' merely asserts that the states in question are mutually accessible: it does *not* imply that $\mathfrak{S}$ and $\mathfrak{S}'$

are necessarily one and the same state. (See also Section 57.)] Again, it is convenient to denote by $\mathscr{S}$ the set of all possible states in which K (not merely K_0) might find itself. Then, the phrase 'if $\mathfrak{S}$ is any arbitrarily prescribed state of the system' may be put more briefly as 'for any $\mathfrak{S}$ in $\mathscr{S}$'. The important property (22.5) of the *relation of adiabatic accessibility*, valid for standard systems, may then be stated in the form

$$\textit{for any } \mathfrak{S}, \mathfrak{S}' \textit{ in } \mathscr{S}, \textit{ if } \mathfrak{S} \nless \mathfrak{S}' \textit{ then } \mathfrak{S}' < \mathfrak{S}. \qquad (37.1)$$

The signs $<$ and $\nless$ when placed between symbols representing states have precisely the meanings just assigned to them, and they must not be thought of as having any arithmetical connotations. Indeed, in some contexts it may be desirable to use in place of $<$ another symbol such as $\prec$; but for the time being this need be of no concern. It may be noted that the accessibility relation has the properties of being (i) *reflexive*, since for any $\mathfrak{S}$ in $\mathscr{S}$ $\mathfrak{S} < \mathfrak{S}$; and (ii) *transitive*, since for any $\mathfrak{S}$, $\mathfrak{S}'$, $\mathfrak{S}''$ in $\mathscr{S}$, $\mathfrak{S} < \mathfrak{S}'$ and $\mathfrak{S}' < \mathfrak{S}''$ entails $\mathfrak{S} < \mathfrak{S}''$. On the other hand, it is non-symmetric since $\mathfrak{S} < \mathfrak{S}'$ and $\mathfrak{S}' < \mathfrak{S}$ may or may not hold simultaneously.

38. The Second Law

(*a*) In the preceding section it was already remarked that if $\mathfrak{S}$ is any state in $\mathscr{S}$ then there always exist other states $\mathfrak{S}'$ in $\mathscr{S}$ such that K_0 cannot undergo a transition from $\mathfrak{S}$ to $\mathfrak{S}'$. The condition of adiabatic isolation of course plays as crucial a part here as it did in Section 21. Consider a simple, albeit rather trivial example, namely that of a fluid at some temperature t contained in an adiabatic enclosure of volume x. Then it is a matter of experience that it is not possible for this system K_0 to attain a final state in which the volume is again x, but the temperature t' is less than t. Here the limitation upon admissible empirical temperature scales introduced in Section 26 is of course taken for granted. A reduction of the temperature of K_0 requires that the fluid do work on the surroundings, but then its volume will have to increase. The subsequent restoration of the original volume will then induce a rise in temperature. If the whole process was quasi-static, $t' = t$. If it was not, either the increase or the decrease in volume, or both, were non-static. The more quickly the increase of x is effected, the less work is done *by* K_0; and the more rapidly the volume is decreased the greater the work done *on* K_0 will be. In any event therefore the surroundings will have done work on K_0 after the initial volume has been restored: its energy will have increased, and $t' > t$. In short, no final state $\mathfrak{S}'(x, t')$ can be reached which has $t' < t$, however small $t - t'$ may be.

(*b*) If one contemplates more complex systems such as the standard system of Section 10*c* it is not quite so easy to give an analogous sufficiently transparent discussion. However, there also, if $\mathfrak{S}(\bar{x}, t)$ is an initial state no final state $\mathfrak{S}'(\bar{x}, t')$ is accessible for which $t' < t$. Note that $\mathfrak{S}'$ is only required to be *isometric with* $\mathfrak{S}$, i.e. $\bar{x}' = \bar{x}$, the actual transition need not be isometric.

Finally, then, on the basis of the outcome of a large number of experiments on systems of any complexity, one generalizes a common feature possessed by all of them in the

Second Law: in every neighbourhood of any state $\mathfrak{S}$ *of an adiabatically isolated system there are states inaccessible from* $\mathfrak{S}$. (38.1)

This formulation of the Second Law is also known as the 'Principle of Carathéodory'. It is not intended to be limited in its applicability to standard systems though in the first instance attention will be largely confined to these.

Given any state $\mathfrak{S}'$ of K_0 one will naturally expect to be able to assert whether or not some other given state is accessible from $\mathfrak{S}'$. In other words, one may hope to be able to define some new physical quantity, $S(x)$ say, such that if S' and S'' are its values in the states $\mathfrak{S}'$, $\mathfrak{S}''$ respectively, then a knowledge of these numbers alone will make it possible to say definitely whether $\mathfrak{S} < \mathfrak{S}'$, or $\mathfrak{S}' < \mathfrak{S}$, or both. This function, the *entropy of K*, does indeed exist, and it therefore *characterizes the relative adiabatic accessibility of states of K*. Chapters 5 and 6 concern themselves with it at length.

It may be of interest to write (38.1) a little differently, using the notation introduced previously. Thus

For any arbitrarily chosen state $\mathfrak{S}$ *and* $\epsilon > 0$ *there are states* $\mathfrak{S}'$ *such that* $\mathfrak{S} \nless \mathfrak{S}'$ and $d(\mathfrak{S}, \mathfrak{S}') < \epsilon$. (38.2)

That $\mathfrak{S}$, $\mathfrak{S}'$ are in $\mathscr{S}$, and that the system is adiabatically isolated, is of course understood in this. With regard to the distance d between $\mathfrak{S}$ and $\mathfrak{S}'$ the notation differs a little from that used hitherto. A state $\mathfrak{S}$ is a set of values of the coordinates of K. However, one may not wish to commit oneself to any particular set of coordinates. The symbol $d(\mathfrak{S}, \mathfrak{S}')$ conveniently represents the distance function for *any* choice of coordinates: if the latter are $y_1, \ldots, y_n$, for instance, then

$$d(\mathfrak{S}, \mathfrak{S}') \equiv d(y, y') = \left\{ \sum_{k=1}^{n} (y_k - y_k')^2 \right\}^{\frac{1}{2}};$$

and this usage is in harmony with Section 19*d*.

(*c*) It will be observed that although (38.1) certainly declares that amongst the final states $\mathfrak{S}'(x, t')$ of the system considered in the first part of this section there will be some which are inaccessible from $\mathfrak{S}(x, t)$, no matter how small $|t-t'|$ may be, it does not fully reflect the particular conclusions reached there, for it says nothing about the sign of $t-t'$. In other words it does not by any means exclude the possibility that the inaccessible states might be just those which have $t' > t$, whilst all states with $t' < t$ would then be accessible. That the contrary situation in fact obtains in nature is an additional result of experience, and it will be more or less obvious at what stage it has to be *explicitly* introduced as such, depending upon how one proceeds to develop the consequences of the Second Law. However, the additional information is already contained in (22.6), whose validity was accepted earlier. Of course, some other kind of empirical result might be adopted instead such as one used later, namely that which follows immediately after (48.9). At any rate, the point to which attention is drawn here is that strictly speaking (38.1) constitutes a proper formulation of the Second Law *only if* an ancillary law of the kind of (22.6) is understood to be taken jointly with it (see also Section 45*b*).

39. The Second Law and integrability

If the Second Law (38.1) is valid for all adiabatic transitions then it must hold *a fortiori* for quasi-static transitions. The procedure of this chapter, unlike that of Chapter 6, is distinguished chiefly by its concentration upon quasi-static processes until a very late stage of the argument has been reached. Accordingly, contemplate

(38.1) in terms of the representative space R_n. Quasi-static transitions of K_0 are represented by solution curves of (30.5). Let $\mathfrak{S}$ be an arbitrarily selected state in $\mathscr{S}$. Then according to (38.1) there exist in every neighbourhood of its representative point $\mathfrak{S}$ representative points of states $\mathfrak{S}'$ inaccessible from $\mathfrak{S}$ along a solution curve of (30.5). In view of the theorem (36.1) it follows immediately that the equation must be integrable, that is to say, the linear differential form dQ of (30.2) must be integrable. This means, then, that there must exist functions $\lambda(x)$ and $s(x)$ such that identically

$$dQ = \lambda(x)\, ds(x). \tag{39.1}$$

40. The empirical entropy

The result (39.1) which has just been inferred directly from the Second Law shows that all quasi-static adiabatic transitions from some initial state $\mathfrak{S}'$ are represented in R_n by curves through $\mathfrak{S}'$ lying in a fixed hyper-surface, the equation of which is

$$s(x) = \text{const.} = s(x'). \tag{40.1}$$

Any continuous sequence of states, no two members of which are quasi-statically accessible from each other, therefore generates a family of non-intersecting hyper-surfaces in R_n, called the adiabatic hyper-surfaces, or simply the *adiabatics* of K.

The new function $s(x)$ which has now appeared in the theory is called the *empirical entropy* (*function*) of K, and its value for a given state $\mathfrak{S}$ is the empirical entropy of the state $\mathfrak{S}$, or of K in the state $\mathfrak{S}$. A basic property, then, of the empirical entropy function is that *it has the same value for all states which are accessible from each other by quasi-static adiabatic transitions.* On the other hand, if a transition from $\mathfrak{S}'$ to $\mathfrak{S}''$ is merely quasi-static the empirical entropy changes by an amount

$$\Delta s = \int_{x'}^{x''} dQ/\lambda = \int_{x'}^{x''} \sum_{i=1}^{n} \lambda^{-1} X_i\, dx_i. \tag{40.2}$$

The path of integration is immaterial since ds is, by definition, a total differential.

The empirical entropy shares with the empirical temperature

the property of being defined only to within scale, meaning that if a particular function, $s(x)$ has somehow been obtained then

$$s^*(x) = g(s(x)) \tag{40.3}$$

is an equally good empirical entropy function, where $g(s)$ is an arbitrary (differentiable) monotonically increasing or decreasing function of s. To s^* there corresponds of course the integrating denominator

$$\lambda^* = \lambda/g'(s); \tag{40.4}$$

but this is nothing new, having been already considered in Section 33 *d*.

The adiabatics, being hyper-surfaces of constant (empirical) entropy, are also called *isentropics*. When $n = 2$ they degenerate into the familiar isentropic curves. However, just in this case they are in a sense trivial; for every linear differential form in only two variables is integrable, as mentioned in Section 33 *d*. Consequently, even if the Second Law were not valid one would still have isentropics when $n = 2$, though not, of course, when $n > 2$.

41. The absolute temperature

To progress essentially beyond the point reached at the end of Section 39, consider two standard systems K_A, K_B. The coordinates of K_A shall be $x = x_1, \ldots, x_n$ and those of K_B $y = y_1, \ldots, y_m$ where it is understood that both x_n and y_m shall be the empirical temperatures of the systems concerned, i.e. $x_n = t_A$, $y_m = t_B$. On account of (39.1) one has

$$dQ_A = \lambda_A(x)\,ds_A(x), \quad dQ_B = \lambda_B(y)\,ds_B(y) \tag{41.1}$$

for any quasi-static transitions of the two systems. Now let the latter be brought into mutual diathermic contact, a compound standard system K_C being so formed, as usual. Under conditions of equilibrium one must necessarily have $t_A = t_B$ and it suffices to use the single symbol t. A state of K_C is now a set of values of the $n+m-1$ coordinates $\bar{x}, \bar{y}, t\,(=x_1, \ldots, x_{n-1}, y_1, \ldots, y_{m-1}, t)$. However, K being a standard system in its own right, one must have

$$dQ_C = \lambda_C\,ds_C, \tag{41.2}$$

where λ_C and s_C are functions of the coordinates of K_C, i.e. of $\bar{x}, \bar{y}, t$. When K_C changes quasi-statically, then so do K_A and K_B. Accordingly, under conditions such that (24.1) holds one has in view of (24.2) the identity

$$\lambda_C ds_C = \lambda_A ds_A + \lambda_B ds_B. \tag{41.3}$$

So far the work of this section has been rather trivial. To arrive at new conclusions observe first that, s_A and s_B being known functions in principle, one can introduce s_A in place of one of the x_k and s_B in place of one of the y_k, after the fashion of Section 12*a*. Without loss of generality it will be supposed that x_{n-1} and y_{m-1} have been eliminated in this way. λ_A is now a function of $x_1, \ldots, x_{n-2}, s_A, t$; λ_B is a function of $y_1, \ldots, y_{m-2}, s_B, t$; whilst λ_C and s_C are functions of both sets. (The original functional symbols are however retained.) Write (41.3) for the moment as

$$ds_C = (\lambda_A/\lambda_C) ds_A + (\lambda_B/\lambda_C) ds_B. \tag{41.4}$$

Now the right-hand member is the total differential of ds_C, yet the differentials of $x_1, \ldots, x_{n-2}, y_1, \ldots, y_{m-2}, t$ do not appear. It follows that s_C must be independent of all these variables. In that case λ_A/λ_C and λ_B/λ_C cannot depend upon them either. This implies in turn that λ_C must be independent of $x_1, \ldots, x_{n-2}, y_1 \ldots, y_{m-2}$. This may be seen at once by supposing it to depend on y_k, chosen at random. Since λ_A certainly does not depend on y_k, this variable could not then be absent from the ratio λ_A/λ_C, contrary to the previous conclusion. It follows that $\lambda_A, \lambda_B, \lambda_C$ are at most functions of (s_A, t), (s_B, t) and (s_A, s_B, t) respectively. Recalling that λ_A/λ_C and λ_B/λ_C are independent of t, each of the three functions in question can depend upon t only through a *common* factor $T(t)$. Nor can this factor be simply a constant since otherwise λ_A for instance would be a function of s_A only, making dQ_A a total differential, which it is certainly not in general. Accordingly

$$dQ_A = T(t)\phi_A(s_A) ds_A, \quad dQ_B = T(t)\phi_B(s_B) ds_B,$$
$$dQ_C = T(t)\phi_C(s_A, s_B) ds_C. \tag{41.5}$$

The function $T(t)$ is a universal function in the sense that, some empirical temperature scale having been adopted to assign temperatures to a number of different systems $K_A, \ldots$, the *same* function $T(t)$ of t will serve as an integrating denominator of $dQ_A, \ldots$, independently of the specific properties of the systems concerned. $T(t)$ is therefore called the *absolute temperature function*, and in the sense just explained the Second Law thus selects a preferred temperature function from the whole set of functions admitted in Section 16. (See also Sections 45 and 49.)

42. The metrical entropy

If (41.3) be combined with (41.5) one gets

$$\phi_C(s_A, s_B)\, ds_C(s_A, s_B) = \phi_A(s_A)\, ds_A + \phi_B(s_B)\, ds_B. \tag{42.1}$$

The first and second members on the right are simply the total differentials of functions of s_A and s_B respectively, $S_A(s_A)$ and $S_B(s_B)$ say. Just as s_A and s_B were previously introduced as coordinates in place of x_{n-1} and y_{m-1}, so s_A and s_B may in turn be replaced by S_A and S_B. Then, retaining the functional symbols on the left of (42.1), the latter becomes

$$\phi_C(S_A, S_B)\, ds_C(S_A, S_B) = dS_A + dS_B = d(S_A + S_B). \tag{42.2}$$

It follows that s_C and therefore ϕ_C must be functions of $S_A + S_B$ alone, say s_C^* and ϕ_C^*:

$$\phi_C^*(S_A + S_B)\, ds_C^*(S_A + S_B) = d(S_A + S_B). \tag{42.3}$$

If one writes $$S_C = \int \phi_C^*(S_A + S_B)\, ds_C^*(S_A + S_B), \tag{42.4}$$

(42.1) becomes $$dS_C = dS_A + dS_B. \tag{42.5}$$

The result which has now been arrived at is the following. Given some empirical temperature scale t there exists a universal function $T(t)$ such that the linear differential form dQ of any standard system K (its 'heat element') becomes, on division by T, the total differential of a function S called the *metrical entropy* of K; and in a quasi-static transition the change of the metrical entropy of a compound standard system is the sum of the changes of the metrical entropies of its constituent standard systems.

In the remainder of this chapter the term 'entropy' will always be intended to mean the metrical entropy.

43. Additivity of entropy

(*a*) The entropy $$S(x) = \int \phi(s)\, ds \qquad (43.1)$$

is a function of the coordinates of K, say of $\bar{x}$, t; and it is evidently defined only to within an additive constant of integration. One therefore assigns some convenient value to the entropy of some standard state of K and then the entropy of all other states is fixed uniquely. An additive constant of integration is still available in (42.4). This may then be chosen in such a way that (42.5) on integration becomes

$$S_C = S_A + S_B. \qquad (43.2)$$

The entropy of a compound (standard) system is therefore the sum of the entropies of its constituent systems ('additivity of entropy'). The validity of this result is assured if and only if (24.1) holds.

(*b*) So far K_A and K_B have been regarded as jointly constituting a standard system K_C, implying that they were in diathermic contact. The question now arises whether one can consistently assign an entropy to a compound system K_C which is non-standard in the sense that it is made up of standard subsystems which are adiabatically separated from one another. In this situation (43.2) will be taken as a *definition* of S_C. Whether it is satisfactory under all circumstances remains to be seen (Section 46). For the time being it may merely be noted that in any adiabatic quasi-static transition of K_C the transitions of K_A and K_B are likewise adiabatic and quasi-static, so that S_A and S_B are constant. S_C will therefore also be constant in such a transition.

It may be worth remarking on how it comes about that one cannot directly pursue the argument of Section 39 when contemplating a system containing an internal adiabatic partition. The number of coordinates of K_C is now $n+m$, which exceeds by one the number previously required. This observation in itself is not strictly relevant. Thus, as has sometimes been suggested, one might arrange the internal partition to be freely movable. The number of coordinates would then be reduced to $n+m-1$ since in any state of K_C the pressures on either side of the partition would have to be equal. No, the salient point is simply that the condition of adiabatic isolation of K_C here is not simply $dQ_C = 0$, but rather splits up into *two distinct* conditions $dQ_A = dQ_B = 0$. Consequently one does not have merely a single differential equation to deal with, contrary to the situation of Section 39. If one inadvertently overlooks this point one is likely to get involved in all sorts of contradictions.

44. The principle of increase of entropy

(*a*) So far in this chapter only part of the content of the Second Law (38.1) has been exhausted, for the two new physical quantities, metrical entropy S and absolute temperature T, have been defined in the course of arguments relating to quasi-static processes alone. The statement of the Second Law is, however, more general, in as far as it covers adiabatic processes of any kind. It is therefore natural to inquire what further conclusions can be drawn from the consideration of general, i.e. not necessarily quasi-static, transitions.

Accordingly, let $\mathfrak{S}'$ be some arbitrarily chosen initial state of an adiabatically isolated standard system K_0. The question is now how one can characterize in a general way those states $\mathfrak{S}''$ which are accessible from $\mathfrak{S}'$, in terms of quantities already defined. To arrive at an answer to this, choose $\bar{x}$, S as coordinates. Consider transitions from some given state $\mathfrak{S}'(\bar{x}', S')$ to states $\mathfrak{S}''(\bar{x}'', S'')$ the values of the deformation coordinates of which are prescribed. The possible values S'' of S cover some range $\mathscr{R}$. [Recall that the possible final values of the non-deformation coordinate cover some range which depends on the way in which the transition proceeds; for instance, there is the possible action of a stirrer, the deformation coordinates may be adjusted at varying rates, and so on.] It is physically plausible to assume that $\mathscr{R}$ is connected (cf. (22.6)), that is to say, it is to be taken for granted that if $\mathfrak{S}_1(\bar{x}'', S_1)$ and $\mathfrak{S}_2(\bar{x}'', S_2)$ are states accessible from $\mathfrak{S}'$ then so are all states $\mathfrak{S}''(\bar{x}'', S'')$ such that S'' lies between S_1 and S_2.

Now S' must certainly lie in $\mathscr{R}$, since the entropy of every state which can be reached quasi-statically from $\mathfrak{S}'$ has the value S'. One can however say much more than this, namely, that S' must be one of the end-points of $\mathscr{R}$. If, on the contrary, S' were an interior point of $\mathscr{R}$ then the system could undergo transitions to states $\mathfrak{S}''(\bar{x}'', S'')$ such that S'' has *any* value sufficiently close to S'. In that case the deformation coordinates could subsequently be adjusted *quasi-statically* to any desired final values, the entropy remaining constant in this process; so that altogether every state in a certain neighbourhood of $\mathfrak{S}'$ could be reached from $\mathfrak{S}'$. This conclusion is in conflict with the Second Law. The following important result has thus been arrived at: *the states $\mathfrak{S}''$ which can be reached adiabatically from $\mathfrak{S}'$ are such that either $S'' \geqslant S'$ for all of them, or else $S'' \leqslant S'$ for all of them.*

Whichever alternative applies when the initial state is $\mathfrak{S}'$ must apply whatever the initial state may be. To see this it suffices to consider the set of all states having given values of their deformation coordinates in common, in view of the constancy of S in quasi-static transitions. Now suppose for the moment that all such states $\mathfrak{S}''$ accessible from an initial state $\mathfrak{S}_1$ had $S'' \geqslant S_1$, whilst all those accessible from an initial state $\mathfrak{S}_2$ had $S'' \leqslant S_2$. (Of course one must have $S_2 < S_1$, or else all states in a neighbourhood of $\mathfrak{S}_2$ would be accessible from the latter, i.e. via $\mathfrak{S}_1$.) If $\mathfrak{S}^*$ is some state such that $S_2 < S^* < S_1$, it would then follow that $\mathfrak{S}_1 \not< \mathfrak{S}^*$ and $\mathfrak{S}_2 \not< \mathfrak{S}^*$. However, because of (37.1) this entails that $\mathfrak{S}^* < \mathfrak{S}_1$ and $\mathfrak{S}^* < \mathfrak{S}_2$; that is to say, $\mathfrak{S}^*$ is a state from which states of both greater and smaller entropies are attainable, which has already been shown to be impossible.

(*b*) With regard to the actual sign of the entropy change in adiabatic transitions it is obvious that no conclusion can be arrived at until some conventional choice has been made about the sign of a multiplicative constant still hidden in the definition of the entropy and absolute temperature functions. These two functions first turn up jointly in (41.5),

$$dQ = TdS, \tag{44.1}$$

and evidently one may write this equally well as

$$dQ = (aT)d(a^{-1}S), \tag{44.2}$$

where a is a constant which may be chosen at will. This constant will make its appearance in any empirical determination of the functions in question, as will be seen in the example of Section 48. The usual procedure is therefore to leave the sign of the entropy change open until such time as one has shown that ΔS is in fact non-negative provided the sign of a be so chosen that $T > 0$. However, one may invert the order of things by requiring that the sign of a be chosen in such a way that $\Delta S \geqslant 0$. This result now holds for any arbitrarily selected standard system, i.e. one has the

Principle of increase of entropy: the entropy of the final state of any adiabatic transition is never less than that of its initial state. (44.3)

This result is sometimes simply referred to as the 'Entropy Principle'. Its relevance to non-standard systems forms the subject of Section 46.

It is regrettable that the Entropy Principle is sometimes stated as applying to *isolated* systems: this is unnecessarily restrictive for it is only necessary for the validity of the Principle that the system should be *adiabatically* isolated.

45. Sign of the absolute temperature

(*a*) Consider an isometric adiabatic transition from some arbitrary state $\mathfrak{S}$ to some neighbouring state $\mathfrak{S}'$. Suppose that the energy of K_0 has increased by dU, so that the empirical temperature will have increased by an amount dt. In order to calculate the corresponding change in entropy dS one need only allow the system to attain $\mathfrak{S}'$ from $\mathfrak{S}$ quasi-statically, the condition of adiabatic isolation having of course been removed. Then

$$dS = dQ/T = dU/T, \tag{45.1}$$

no work having been done as this transition, too, is supposed to be isometric. Granted that $T(t)$ is continuous, (45.1) shows that $dS \neq 0$. Returning to the adiabatic transition considered initially it follows from (44.3) that $dS > 0$. But $dU > 0$, and therefore

$$T > 0. \tag{45.2}$$

Moreover, T is a single-valued function of t. Thus, since it is a universal function it suffices to take $n = 2$. Anticipating some of the results of Section 48, (48.6) then shows that dT/dt cannot vanish; and reference to the properties of a specific class of substances [equations (48.12, 13)] shows that in fact

$$dT/dt > 0. \tag{45.3}$$

(*b*) At the risk of some repetitiveness let it be remarked that the result (45.2) became available only after accepting some empirical information concerning *possible* non-static transitions, namely that when the latter are isometric work has to be done *on* K_0, so that $\Delta U > 0$. Other possibilities offer themselves, such as that to be considered in Section 48*a*. The need for introducing such information was already discussed in Section 38*c*. It need in any case not come as a surprise, for on the one hand S and T were defined in this chapter on the basis of properties of quasi-static transitions alone, whilst on the other hand the Second Law in the form (38.1) does not explicitly say what kind of states are adiabatically inaccessible, but merely that such states exist. If one is satisfied with stating the Second Law *after* energy has been defined one might therefore introduce U as the non-deformation coordinate from the outset and then declare, in place of (38.2), that if K_0 is any adiabatically isolated system then

For any arbitrarily chosen state $\mathfrak{S}(\bar{x}, U)$ *and* $\epsilon > 0$ *there are states* $\mathfrak{S}'(\bar{x}, U')$ *such that* $\mathfrak{S} \nprec \mathfrak{S}'$ *and* $-\epsilon < U' - U < 0$. (45.4)

Granted the usual continuity assumptions this will then allow one not only to define S and T but also to arrive at (44.3) and (45.2) (see also Section 50*a*).

(*c*) However this may be, the possibility of deducing (45.2) hinged also on the assumption that there existed in fact quasi-static (non-adiabatic) transitions between the terminal states of the irreversible transitions introduced at the beginning of this section. *A priori* it is not necessary that this should always be so. 'Systems' have upon occasions been considered for which the assumption is false (nuclear spin systems), and for these both positive and negative absolute temperatures have to be admitted. Equation (45.4) is then also invalid for a whole class of states, though (38.2) and the Principle of Increase of Entropy may be retained. Negative temperatures are to be considered as hotter than any positive temperature in the sense of Section 26, i.e. when two systems one of which has $T < 0$ whilst the other has $T > 0$ are brought into diathermic contact heat will flow from the first to the second. Still, though a sensible thermodynamic description of such systems appears to be possible one may be somewhat reluctant to fit them into the framework of phenomenological thermodynamics. Thus, for instance, the nuclei of the atoms whose spins constitute the 'system' under discussion ($T_s < 0$) form a second system ($T_l > 0$) and these are of necessity coupled together. This coupling represents diathermic contact which is outside one's control. Spin systems can therefore in practice be thought of as adiabatically isolated only in a limited sense, and certainly only if the spin-lattice coupling is very weak. Their states are thus somewhat ephemeral, and they will not be further considered.

46. Entropy principle for non-standard systems

(*a*) The Principle of Increase of Entropy (44.3) has so far been shown to be valid for standard systems only. What therefore is its status as regards: (i) systems which are not standard on account of the presence of internal adiabatic partitions; and (ii) processes resulting from the thermal coupling of standard systems which are initially at different temperatures?

First, let K_C consist, as usual, of two subsystems K_A and K_B, adiabatically separated from each other. In an adiabatic transition of K_C as a whole, the transitions of K_A and K_B are both adiabatic so that $\Delta S_A \geqslant 0$ and $\Delta S_B \geqslant 0$. Adopting the definition $S_C = S_A + S_B$ for the entropy of K_C, as tentatively accepted in Section 43*b*, it follows that

$$\Delta S_C \geqslant 0. \tag{46.1}$$

Next, let K_A and K_B be adiabatically separated as above, and $T'_A > T'_B$ initially. Let the condition of mutual adiabatic isolation then be relaxed. When K_C has attained equilibrium the common temperature shall be T''. The results of Section 26, taken together with (45.3), imply

$$T'_B < T'' < T'_A. \tag{46.2}$$

The transition of K_C as a whole is adiabatic and is supposed to take place isometrically. The heat Q gained by K_B is exactly that lost by K_A,

$$Q_B = -Q_A = Q. \tag{46.3}$$

The changes of entropy of K_A and K_B may be calculated after the manner of Section 45*a*, i.e. by imagining their final states to be separately attained by quasi-static (diathermic) processes. These are to be carried out in such a way that T_A and T_B change monotonically, so that $dQ_A < 0$ and $dQ_B > 0$. Then in view of the first of these inequalities and since $T_A > T'' > 0$

$$\Delta S_A = \int dQ_A/T_A > (1/T'') \int dQ_A = -Q/T'',$$

and similarly, since $T'' > T_B > 0$ and $dQ_B > 0$,

$$\Delta S_B = \int dQ_B/T_B > (1/T'') \int dQ_B = Q/T''.$$

Hence, by addition, $$\Delta S_C > 0. \tag{46.4}$$

Thus the general principle $\Delta S \geqslant 0$ is again seen to be satisfied, indeed strongly so, in the sense that the possibility of S_C remaining constant is excluded in this case. The transition is of course irreversible; for in the process of trying to restore the initial temperature difference, the internal adiabatic partition must be restored at some stage. Prior to this K_C was a standard system, and its entropy cannot have decreased. In the second part of the process the entropy of K_C would therefore have to decrease by an amount ΔS_C at least. This, however, is impossible since K_A and K_B are now separately changing adiabatically.

So far the arguments of this section have been intended merely to serve as an introduction. It is not difficult to convince oneself that by suitably generalizing them one will be led to analogous conclusions in situations differing somewhat from those which have been explicitly described. What is being contemplated now are processes involving the (temporary) substitution of diathermic for adiabatic internal partitions, together with simultaneous changes of the deformation coordinates. (Initially the internal partition must of course be adiabatic.) Indeed, let a state of K_C be regarded as a simultaneous state of K_A and K_B, so that the initial state, for instance, is $\mathfrak{S}'(\bar{x}', \bar{y}', S'_A, S'_B)$. If the partition happens to be diathermic at some stage, one of the coordinates will be redundant but this is of no account. Let an adiabatic transition take K_C to $\mathfrak{S}''(\bar{x}'', \bar{y}'', S''_A, S''_B)$, and suppose that

$$S''_C - S'_C \equiv -\delta < 0. \tag{46.5}$$

Next, let K_C go from $\mathfrak{S}''$ to a state $\mathfrak{S}^*(\bar{x}', \bar{y}', S_A^*, S_B^*)$ quasi-statically, this transition being so arranged that $S_A^* = S_A' - \sigma$, where σ is some sufficiently small positive quantity. [This is always possible if the partition between K_A and K_B be suitably manipulated. Thus one may go first to a state such that S_A has the required values S_A^*, whilst S_B is allowed to take care of itself, the partition being diathermic. Thereafter, whilst it is adiabatic, the deformation coordinates are adjusted as required.] At any rate, $S_C'' = S_C^*$, whence

$$S_B^* = S_B' + \sigma - \delta.$$

Hence, choosing σ sufficiently small, one has $S_A^* < S_A'$ *and* $S_B^* < S_B'$. Now, however, S_A and S_B may be separately increased by means of non-static adiabatic processes as usual. Keeping in mind that $\bar{x}$ and $\bar{y}$ may be adjusted reversibly at will without S_A and S_B being affected, it follows that all states in a neighbourhood of $\mathfrak{S}'$ are accessible from $\mathfrak{S}'$, a result which conflicts with the Second Law. The assumption (46.5) that the entropy of K_C could have decreased in an adiabatic transition is therefore false. The Entropy Principle (44.3) thus continues to hold under the more general circumstances now contemplated; and when the entropy is known actually to have increased in an adiabatic transition the latter is irreversible. (Reference may also be made to Section 51.)

(*b*) Evidently the argument leading to the Principle of Increase of Entropy is more straightforward for standard systems than for non-standard systems. The main reason for this is that in the latter case one is contemplating processes which involve *modifications of the given system* (cf. Section 14*c*). In the example of the thermal conduction considered above, one is strictly speaking considering one system K_C prior to the removal of the adiabatic partition, but another system K_C after it. (One sees this quite clearly reflected in the change of the number of coordinates of K_C consequent upon the removal of the partition.) If, therefore, reference has been made to the change of entropy of 'the system K_C' this was possible only in virtue of the agreement, left understood, to refer to the collection of bodies which went to make up K_C as 'the system' at all times. This point of view brings out very clearly the function of (43.2) in this case as the *definition* of S_C. Evidently the mere process of inserting an adiabatic partition within a standard system in equilibrium does not affect the *value* of the entropy although it requires that one go over to a new entropy *function.*

47. Perpetual motion of the second kind

(*a*) Recall the general remarks of Section 28 concerning perpetual motion machines. There is was pointed out that one was of course concerned with *useful* motion. This meant in effect that one required some system which in the course of a closed adiabatic transition could do a positive amount of work on its surroundings. The First Law immediately declares this to be impossible.

Now, attempts were also made in the past to construct so-called 'perpetual motion machines of the second kind'. Again the motion is to be useful. In this case one seeks a 'periodic' device K_B for extracting energy from a system K_A, such that after one period K_B is in its initial condition whilst the only change in K_A is a decrease of its temperature. The compound system K_C consisting of K_A and K_B as a whole can only interact mechanically with its surroundings. No such arrangement is in fact possible, for the entropy of K_C at the end of a period would have to be less than its entropy at the beginning of the period, in contravention of the Entropy Principle.

The Second Law thus implies the non-existence of perpetual motion machines of the second kind; and the persistent failure of all attempts to construct them is just evidence for the validity of the Second Law. It may also be noticed that to some extent the term 'perpetual motion' is altogether a misnomer here. For even if one could construct the kind of device in question the amount of energy which could thus be obtained is finite, in accordance with the remarks at the end of Section 25*b*.

(*b*) Recalling the discussion of Section 29, one becomes aware that the Second Law, and the Entropy Principle, fit naturally into its framework. The Second Law is stated directly in the form of a law of impotence, as is its corollary—the impossibility of the perpetual motion of the second kind. With regard to the Entropy Principle, on the other hand, one is confronted with a rare, perhaps a unique, example of a *one-sided* conservation law. Whereas in adiabatic transitions the energy of a system could neither increase nor decrease, its entropy is under like conditions prohibited only from decreasing. In other words, entropy cannot be destroyed, but it can very well be created.

48. Empirical determination of S and T

(*a*) The energy, entropy and absolute temperature of a system may be determined experimentally in various ways. When the system is complex one will often be able to reduce the problem to that of determining the entropies of simple standard subsystems separately, in view of the additivity of energy and entropy. One may with profit therefore investigate how to proceed when $n = 2$. A very instructive procedure—though not necessarily the most convenient in practice—is to begin with the determination of empirical temperature and entropy functions. Thus one finds (i) the isothermals, essentially in the way described at the end of Section 18; and (ii) the isentropics, by allowing the system K_0 to undergo quasi-static changes from a sequence of initial states which cannot be linked quasi-statically. In this part of the experiment one is merely making use of the integrability of dQ, i.e. of the mere existence of isentropics. It suffices to take K as a fluid exerting a pressure P on its enclosure whose volume is V. Then the equation of any isothermal is

$$t(P, V) = \text{const.,} \tag{48.1}$$

and that of any isentropic is

$$s(P, V) = \text{const.} \tag{48.2}$$

The functions t and s are known explicitly, some definite way of labelling the various curves having been adopted. These functions may be taken as the empirical temperature and empirical entropy functions of K. The absolute temperature T and metrical entropy S are as yet unknown functions of these, say $\tau(t(P, V))$ and $\sigma(s(P, V))$ respectively. Now

$$TdS = dU + PdV, \tag{48.3}$$

whence $$\tau\sigma'(s_{,P}dP + s_{,V}dV) = U_{,P}dP + (U_{,V} + P)dV. \tag{48.4}$$

Primes indicate derivatives with respect to the argument where functions of one argument are concerned. On the other hand, a variant of the notation explained in Section 33*a* applies to partial derivatives: if F is a function of variables A, B,..., then its partial derivatives with respect to A, B,... are written $F_{,A}, F_{,B}$,.... The second partial derivatives are naturally denoted by $F_{,AA}, F_{,AB}$,.... If the function to be differentiated already has a subscript this causes no worry as it will occur before the (subscript) comma; e.g. $\partial^2 F_K/\partial A \partial B = F_{K,AB}$. The use of this notation very greatly improves the appearance of thermodynamic equations.

Now in (48.4) dP and dV are independent, whence

$$U_{,P} = \tau\sigma' s_{,P}, \quad U_{,V} = \tau\sigma' s_{,V} - P. \tag{48.5}$$

Differentiating the first of these with respect to V and the second with respect to P, the identity $U_{,PV} = U_{,VP}$ gives rise to

$$\tau'(t)\sigma'(s)\,(t_{,P}s_{,V} - t_{,V}s_{,P}) = 1. \tag{48.6}$$

The expression in parentheses must therefore be the product of functions of (the functions) s and t only. Except for a multiplicative constant a in τ' and its reciprocal $1/a$ in σ', these derivatives may therefore be read off by inspection. Their integrals involve additive constants of integration. That occurring in S remains arbitrary, but that occurring in T can be determined by subsequently considering non-static transitions.

For the sake of illustration let it be supposed specifically that

$$t(P, V) = PV, \quad s(P, V) = PV^{\gamma}, \tag{48.7}$$

where γ (> 1) is a constant. It is known that there exists a class of substances, i.e. the rare gases, for which these assumptions are very nearly satisfied under appropriate conditions. It is only necessary that V and PV should be sufficiently large. Using (48.7) in (48.6) one gets at once

$$t_{,P}s_{,V} - t_{,V}s_{,P} = (\gamma - 1)PV^{\gamma} = (\gamma - 1)s,$$

so that

$$(\gamma - 1)s\tau'(t)\sigma'(s) = 1.$$

It follows, if a is some constant, that

$$\tau'(t) = a, \quad \sigma'(s) = [(\gamma - 1)as]^{-1},$$

or

$$T = at + T_1, \quad S = [(\gamma - 1)a]^{-1}\ln s + S_1, \tag{48.8}$$

where S_1 and T_1 are constants of integration. The energy function may then be obtained from (48.3) by integration. Using (48.7, 8) one gets after a little manipulation

$$U = (\gamma - 1)^{-1}[t + (T_1/a)\ln s] + U_1, \tag{48.9}$$

where U_1 is another constant of integration.

Going on to the information supplied by a non-static experiment it is known that if such an adiabatically enclosed gas be expanded suddenly, no work being done in this process, then t remains constant. On the other hand, $s(= tV^{\gamma-1})$ has obviously increased, and consistency with the constancy of U then demands that

$$T_1 = 0. \tag{48.10}$$

Furthermore, the metrical entropy has increased, which means that the constant a must be taken as positive (cf. the remarks following (44.2)). Then $T > 0$, since P and V cannot be negative; a conclusion which had already been reached in Section 45*a*.

In Section 26 the symbol c was used as an abbreviation for the derivative $U_{,t}(\bar{x} = \text{const.})$. The arbitrariness in the definition of c arising from the arbitrariness inherent in the empirical temperature t can be circumvented by understanding t to have been replaced by the absolute temperature:

$$c = U_{,T}. \tag{48.11}$$

(c is usually known as the 'specific heat at constant deformation coordinates'.) (48.9, 10) show that U depends here on T alone, and

$$c = [a(\gamma - 1)]^{-1}. \tag{48.12}$$

Thus finally

$$\left.\begin{aligned} T &= at = aPV, \\ S = c\ln s + S_1 &= c(\ln P + \gamma \ln V) + S_1 \\ &= c\ln T + a^{-1}\ln V + (S_1 - c\ln a \\ U &= cT + U_1. \end{aligned}\right\} \tag{48.13}$$

(*b*) In practice one might well proceed by determining the energy directly, in place of the empirical entropy. Thus, retaining the first member of the (48.7) in the example above, a direct measurement would reveal that in adiabatic processes of any kind the work done by K_0 is just proportional to the decrease of the produce PV. In other words,

$$U = bt + U_1, \tag{48.14}$$

where b is a positive constant determined by the experiment, and U_1 an arbitrary constant. Then

$$T(t)\,dS = b\,dt + t\,dV/V, \tag{48.15}$$

which shows that t/T must be a constant, say

$$T(t) = at, \tag{48.16}$$

whence immediately

$$S = (b/a)\ln T + (1/a)\ln V + \text{const.}, \tag{48.17}$$

in agreement with (48.12). Lest it seem surprising that the unknown constant T_1 never appears now, observe that (48.14) was intended to apply to all adiabatic processes. If, however, one restricted these to be quasi-static in the first place, then experiment would still show that (48.14) was valid in the sense that

$$\Delta U = b\Delta t,$$

where b is the same positive constant for every initial state, but the undetermined additive constant U_1 could have a different value for every isentropic, and (48.15) would involve further terms.

49. Concerning the definition of absolute temperature

(*a*) It has already been explained in Section 41 that the absolute temperature function is in fact designated 'absolute' for the following reason. Given a system K_A, let its isothermals be

$$f_A(x) = t. \tag{49.1}$$

K_A may be used as a thermometer to determine the isothermals of other systems $K_B, K_C, \ldots$:

$$f_B(y) = t, \quad f_C(z) = t, \ldots. \tag{49.2}$$

Then there exists a function $T(t)$ of t, unique to within a multiplicative constant, such that T is an integrating denominator of $dQ_A, dQ_B, dQ_C, \ldots$; that is to say, a total differential results when $dQ_A = \Sigma X_k(x)\,dx_k$ is divided by the function $T(f_A(x))$, and likewise when $dQ_B = \Sigma Y_k(y)\,dy_k$ is divided by the function $T(f_B(y))$; and so on.

The results derived from (48.6) must therefore of course be independent of the initial choice of empirical temperature and entropy scales. In other words they must not be affected if in place of (48.1, 2) one goes over to

$$t^*(P,V) = f(t(P,V)) = \text{const.},$$
$$s^*(P,V) = g(s(P,V)) = \text{const.}, \qquad (49.3)$$

respectively, with f and g arbitrary. That this is so may be shown without difficulty. Thus if $T = \tau^*(t^*)$, and $S = \sigma^*(s^*)$ one needs to show that

$$\tau^*(t^*) = \tau(t), \quad \sigma^*(s^*) = \sigma(s), \qquad (49.4)$$

whence
$$\frac{d\tau}{dt} = \frac{d\tau^*}{dt^*}\frac{df}{dt}, \quad \frac{d\sigma}{ds} = \frac{d\sigma^*}{ds^*}\frac{dg}{ds}.$$

With these, (48.6) leads to an equation of exactly the same form except that σ, τ, t, s, are replaced by the corresponding starred symbols. This, however, is the equation one would have got if one had started with t^* and s^* instead of t and s in the first place.

The magnitude of the constant a in (44.2), or in (48.13), remains undetermined. It may be fixed by arbitrarily assigning a value to some easily reproducible state of a convenient system, e.g. the value 273·16 to the state of a mixture of ice, water and water vapour in mutual equilibrium.

(*b*) A simple way of constructing a thermometer which reads absolute temperatures directly is to use as thermometric substance a gas which obeys the first member of (48.13) sufficiently closely. However, it seems desirable to enlarge a little on this remark. So-called 'ideal gases' are sometimes defined as gases which obey (i) Boyle's Law

$$PV = \text{const.}\ (t = \text{const.}), \qquad (49.5)$$

and (ii) Charles's Law $\quad V = V_0(1+\alpha t). \qquad (49.6)$

The first of these states that, under suitable conditions, the isothermals of some gases are experimentally found to be rectangular hyperbolas in R_2; the second states that, when P is kept constant. these gases expand linearly with t, the constant α having the same value in each case, independently of P and of the nature of the gas. t is defined in terms of the readings of a mercury-in-glass thermometer of uniform bore. (49.5, 6) operate as *definitions* when they are used as criteria for deciding whether a given gas is to be called 'ideal' or not.

Equations (49.5) and (49.6) may be replaced jointly by the single relation

$$PV = \beta(1+\alpha t), \tag{49.7}$$

where β is a constant; so that according to the definition above a gas is ideal if it obeys the equation of state (49.7). At this stage it is then not unusual to find a new empirical temperature T^* introduced, viz.

$$T^* = 1+\alpha t, \tag{49.8}$$

called the 'gas-temperature'. The latter is thus *defined* by (49.7) with the understanding, of course, that the gas has indeed the isothermals (49.5). Clearly Charles's Law now operates as a definition of a particular empirical temperature scale, although its law-like character is retained to the extent that (49.8) is intended to imply that the same value of α applies to a whole class of gases.

Now what is sometimes not brought out sufficiently clearly is that the equality of the gas temperature T^* and of the absolute temperature T is by no means assured as yet. This is not surprising when one considers that there is an arbitrary element in the choice of T^*; and, further, that nothing has been said yet about the isentropics of the gas. A specific illustration may not be out of place. Write $\beta T^* = t$ so that

$$PV = t, \tag{49.9}$$

and suppose it has been ascertained experimentally that the energy of the 'gas' is

$$U = f(t)+\nu t \ln V, \tag{49.10}$$

where f is some function of t, and ν some constant. Proceeding as in Section 48b, the integrability condition on dS gives

$$(1+\nu)t\frac{dT}{dt} = T,$$

whence

$$T = \text{const.}\, t^{\nu'}, \tag{49.11}$$

where

$$\nu' = (1+\nu)^{-1}.$$

Thus the product PV does not measure the absolute temperature unless $\nu = 0$; in particular, if $\nu = -\frac{1}{2}$ it measures its square root. Generally, for the gas scale (49.9) defined by a substance obeying Boyle's Law to be identical with the absolute temperature it is necessary and sufficient that the energy be a function of t alone. (In Section 48 the form of the isentropics given by (48.7) in fact required this function to be linear; see also Section 71*a*.)

50. Other formulations of the Second Law

(*a*) It may be of interest to refer briefly to two formulations of the Second Law which are generally to be found in treatises on thermodynamics. They are:

(i) *The Principle of Kelvin: it is impossible to construct an engine which, operating in a cycle, will produce no effect other than the extraction of heat from a reservoir and the performance of an equivalent amount of work*; (50.1)

(ii) *The Principle of Clausius: it is impossible to construct a device which, operating in a cycle, will produce no effect other than the transfer of heat from a cooler to a hotter body.* (50.2)

It should be appreciated that the differences between the formulations (38.1), (50.1) and (50.2) are less than superficial inspection would lead one to believe, especially when one recalls the fact that the first of these quietly leaves the need for some ancillary statement understood (cf. Section 38*c*). Indeed, when the latter is taken into account from the start one will have a law of the kind (45.4); but this then comes very close to (50.1). Furthermore, (50.1) amounts simply to the prohibition of perpetual motion machines of the second kind, already contained in Section 47*a*; though here it now functions as a *primary* law. Similarly, (50.2) is already contained in effect in Section 46*a*.

It is largely a matter of taste as to which formulation one prefers. The virtues of that of Carathéodory are, on the one hand, its lack of verbal emphasis on any engineering terms such as 'engine' and 'machine operating in cycles' whilst, on the other hand, it allows more easily of a clear separation of the mathematical from the physical content of the theory. It achieves this by explicitly dealing with the purely mathematical aspects of integrability first, in the form of the Theorem of Carathéodory; whereas if one proceeds along more traditional lines this integrability turns up as a by-product of arguments involving Carnot engines and Carnot cycles, whose function is often further obscured by the insistence on the availability of 'ideal gases'. There is no doubt that part of the difficulty of the 'classical' arguments lies in the subtlety with which mathematical notions and ostensibly physical notions are almost inextricably interwoven.

(*b*) Once the functions S and T are available along with the Entropy Principle all the well-known results concerning the 'efficiency' of engines, the availability of energy, and so on, are of course obtainable without much ado. By way of example, consider a closed transition of a (standard) system K_A, supposing it for the time being to be pseudo-static. To account for the interactions of K_A with its surroundings one may imagine the latter to be equivalent to a system K_B in diathermic contact with K_A, the compound system K_C being isolated. Then

$$\Delta S_C \geqslant 0. \tag{50.3}$$

However, K_A has regained its initial state, so that $\Delta S_A = 0$ and so

$$\Delta S_B \geqslant 0. \tag{50.4}$$

In principle nothing prevents the transition of K_B from being quasi-static, so that

$$\Delta S_B = \oint dQ_B/T_B \geqslant 0. \tag{50.5}$$

However, here $T_B = T_A$ and $dQ_B = -dQ_A$, so that

$$\oint dQ_A/T_A \leqslant 0. \tag{50.6}$$

This is the *Inequality of Clausius*. If one states it uncritically as being valid for *any* closed transition of K_A, as is sometimes done, one runs into difficulties. T_A is then not defined in general; and in (50.6) one must certainly replace T_A by T_B and further suppose T_B to be defined. (In practice this will require 'temperature gradients' within K_B to be negligible.) Of course, no difficulty arises if the transition is pseudo-static in part, whilst the remainder is non-static but adiabatic.

Let (50.6) be applied to the situation in which K_A is used to withdraw an amount of heat Q_+ from a system K_+ so large that its temperature T_+ is not sensibly affected, and an amount of heat Q_- is similarly transferred by K_A to another system K_- at temperature T_-. There are no thermal interactions other than those already specified, so that the work done by K_A in the closed transition ($\Delta U_A = 0$) is $W = Q_+ - Q_-$. (50.6) gives at once

$$Q_+/T_+ - Q_-/T_- \leqslant 0,$$

whence

$$W \leqslant Q_+(1 - T_-/T_+). \tag{50.7}$$

The efficiency $\eta = W/Q_+$ is therefore subject to the inequality

$$\eta \leqslant 1 - T_-/T_+. \tag{50.8}$$

Thus only a certain fraction of the heat Q_+ withdrawn from a system can be converted into work, and for this to be possible at all, another system at a lower temperature must be available. In this context, let there be a third system K_1 at temperature T_1, $T_1 < T_- < T_+$. Then if an amount of heat Q be transferred from K_+ to K_1 the greatest amount of work which can be done is

$$W_+ = Q(1 - T_1/T_+).$$

Suppose this same amount of heat is first transferred by conduction from K_+ to K_-, and then from K_- to K_1: then work can be done only in the second stage of the process, of amount not greater than

$$W_- = Q(1 - T_1/T_-).$$

Evidently the amount of energy which has under these circumstances become unavailable as the result of conduction is

$$W_+ - W_- = T_1 Q(1/T_- - 1/T_+). \tag{50.9}$$

Recalling Section 46, the change of entropy of the system composed of K_+ and K_- consequent upon the transfer of Q between them is just

$$S = Q(1/T_- - 1/T_+),$$

so that

$$W_+ - W_- = T_1 \Delta S. \tag{50.10}$$

The increase in entropy is thus reflected in a decrease of *available* energy. The actual amount of this however does not depend upon ΔS alone.

Finally, as a corollary to (50.6), let the transition contemplated there consist of two parts, from $\mathfrak{S}'$ to $\mathfrak{S}''$ and back from $\mathfrak{S}''$ to $\mathfrak{S}'$, the second of these being quasi-static. Then

$$\int dQ/T + S' - S'' \leqslant 0. \tag{50.11}$$

But $S'' - S' = \Delta S$ is the difference between the entropy of the terminal states of the first part of the transition (to which the integral in (50.11) refers). Hence

$$\Delta S \geqslant \int dQ/T. \tag{50.12}$$

This, then, is valid for any transition such that a meaning can be attached to T, as previously observed.

CHAPTER 6

THE SECOND LAW (II)

51. The ordering of states

The Second Law (38.1) directly expresses a limitation on the relative adiabatic accessibility of states. It implies that if two states $\mathfrak{S}$ and $\mathfrak{S}'$ of a system K_0 be arbitrarily assigned then it may be that no transitions from $\mathfrak{S}$ to $\mathfrak{S}'$ are possible, even though no violation of the laws of mechanics and of the First Law would be involved in such transitions. More precisely, it asserts that in every neighbourhood of $\mathfrak{S}$ inaccessible states do in fact exist:

For any state $\mathfrak{S}$ and any $\epsilon > 0$ there exist states $\mathfrak{S}'$ such that $\mathfrak{S} \nless \mathfrak{S}'$ and $d(\mathfrak{S}, \mathfrak{S}') < \epsilon$. (51.1)

As stressed previously (Sections 38*c* and 45*b*), although (51.1) allows one to define the metrical entropy S, granted the First Law, it does not imply the fully fledged Entropy Principle (44.3). This is not surprising since (51.1) does not say *what kind* of states are inaccessible. In this regard it differs from the traditional formulations (Section 50*a*); and the impression that these are less 'economical' than (51.1) is to some extent illusory. For the time being the existence of the energy function U, with the properties discussed in Chapter 3, will be presupposed. (51.1) will then be supplemented by (22.6), it being understood that only standard systems are now contemplated. Thus

If $\mathfrak{S}(\bar{x}, U)$, $\mathfrak{S}'(\bar{x}, U')$ are states of K_0 isometric with each other then $\mathfrak{S}(\bar{x}, U) < \mathfrak{S}'(\bar{x}, U')$ if $U' \geqslant U$. (51.2)

Note that (51.2) implies the weaker assumption (22.5), i.e.

If $\mathfrak{S}$, $\mathfrak{S}'$ are states of K_0 and $\mathfrak{S} \nless \mathfrak{S}'$ then $\mathfrak{S}' < \mathfrak{S}$, (51.3)

as was shown in Section 22*b*.

Now in Section 9*a* attention was already drawn to the fact that (51.1) implies a certain order in time of states of K_0. Thus, if $\mathfrak{S}$ and $\mathfrak{S}'$ are any two states, their relative accessibility depends upon them alone. When they stand in the relation $\mathfrak{S} \nless \mathfrak{S}'$ then $\mathfrak{S}$ is a *later* state than $\mathfrak{S}'$, in the sense that if K_0 was in one of these states at a certain time and in the other at another time, then the later state must have been $\mathfrak{S}$. The Entropy Principle is just an expression of this order in time, and its basic significance might make a line of development appear natural in which attention is focused on the ordering of states from the outset. In other words, reversible and irreversible transitions are considered jointly from the beginning, in contrast with the procedure of Chapter 5, where only reversible transitions were admitted at first; and where the possibility of arriving at any interesting conclusions hinged on the availability of the differential equation $dQ = 0$.

The present chapter, then, deals with the Second Law from the point of view of the ordering of states. Even under this general heading several different approaches suggest themselves and some of these will be dealt with in turn. They differ largely in the extent to which they presuppose consequences of the First Law. What is common to them is the intention to define, in as direct a way as possible, a continuous (numerical-valued) function $s(x)$ which reflects the ordering under discussion in the sense that if s', s'' are the values of s for the states $\mathfrak{S}'$, $\mathfrak{S}''$ respectively then $\mathfrak{S}' \nless \mathfrak{S}''$ *if and only if* $s'' < s'$.

The conceptual advantages of such a procedure may be brought out by considering a certain finite (or at least denumerable) set $\mathscr{S}^*$ of states. Select any state in $\mathscr{S}^*$, say $\mathfrak{S}_1$. Determine another state $\mathfrak{S}_2$ such that $\mathfrak{S}_1 \nless \mathfrak{S}_2$—according to (51.1) every neighbourhood of $\mathfrak{S}_1$ contains a state of the required kind. Then continue to select states $\mathfrak{S}_3, \mathfrak{S}_4, \ldots, \mathfrak{S}_k$, such that for any j, either $\mathfrak{S}_{j-1} \nless \mathfrak{S}_j$ or $\mathfrak{S}_{j-1} = \mathfrak{S}_j$, as the case may be. Next associate with $\mathfrak{S}_j$ $(j = 1, 2, \ldots)$ a (real) number s_j subject only to the condition that $s_j < s_{j-1}$ if $\mathfrak{S}_{j-1} \nless \mathfrak{S}_j$ and $s_j = s_{j-1}$ if $\mathfrak{S}_{j-1} = \mathfrak{S}_j$. Then it is sensible to think of $s_j = \psi(\mathfrak{S}_j)$ as the empirical entropy of $\mathfrak{S}_j$, and of ψ as an empirical entropy function defined on $\mathscr{S}^*$ $(= \mathfrak{S}_1, \mathfrak{S}_2, \ldots, \mathfrak{S}_k, \ldots)$; for the whole construction has been so designed that $\mathfrak{S}_i \nless \mathfrak{S}_j$ if and only if $s_j < s_i$. In particular if $s_i = s_j$ then $\mathfrak{S}_i = \mathfrak{S}_j$. Otherwise expressed, *in any transition of K_0 between states of $\mathscr{S}^*$ the empirical entropy of the final state is never less than that of the initial state.* This is very much like the Entropy Principle, except that, unfortunately, it relates only to $\mathscr{S}^*$.

Despite the comparative triviality of the remarks relating to what was called the empirical entropy (function) s_j they do shed light on the *raison-d'être* for this function in the general case. Moreover, it is quite clear that one has strictly speaking at this stage only a Principle of Unidirectional Change of Entropy, for in place of s_j one could equally well adopt a labelling $s_j^* = g(s_j)$, where g is any monotonically increasing or decreasing function of s_j. The ordering (or technically speaking the quasi-ordering) of the states would not be affected. It is also of some interest to observe that the argument is concerned with reversible and irreversible transitions rather than with transitions which are quasi-static or non-static.

Leaving $\mathscr{S}^*$ aside now, the situation in the general case is not quite so simple as regards detail, on account of the fact that $\mathscr{S}$ is non-denumerable. One might of course content oneself (in physics!) with simply *postulating* the existence of a continuous (differentiable) empirical entropy function $s(x)$ with the property that, for any states $\mathfrak{S}'$ and $\mathfrak{S}''$, $\mathfrak{S}'$ is adiabatically inaccessible from $\mathfrak{S}''$ if and only if $s(x') < s(x'')$. The second Law in the form (51.1), as well as (51.2), are then redundant, the proposed postulate being a substitute for them. (It is not implied that it is equivalent to (51.1, 3); cf. particularly Section 57.) However this may be, it is no doubt of interest to examine how (51.1), together with whatever further assumptions may have to be granted, will lead to the conclusion that an empirical entropy function with the required properties does indeed exist.

52. Existence of $s(x)$ when the First Law is presupposed

(*a*) For the present the First Law and the possibility of doing an arbitrary amount of work on an adiabatically isolated system K_0 are presupposed, i.e. (51.2) is accepted. Let the energy U be chosen as non-deformation coordinate: $x_n = U$; and let $\mathfrak{S}'$ be an arbitrarily chosen state. In R_n the line $\mathfrak{L}$ through $\mathfrak{S}'$ parallel to the U-axis is effectively the same as that already considered at the beginning of Section 26. If U' is the energy of $\mathfrak{S}'$ then all states $\mathfrak{S}''$ on $\mathfrak{L}$ which have $U'' > U'$ are accessible from $\mathfrak{S}'$, in view of (51.2). Then it is first to be shown that as a consequence of (51.1) all states $\mathfrak{S}''$ on $\mathfrak{L}$ which have $U'' < U'$ are inaccessible from $\mathfrak{S}'$.

[Only adiabatic transitions are contempated throughout.] Suppose therefore that, on the contrary, there exists at least one state $\mathfrak{S}''(\bar{x}, U'')$, $U'' < U'$, such that $\mathfrak{S}' < \mathfrak{S}''$. Then certainly also all states on $\mathfrak{L}$ whose energy exceeds U'' are accessible from $\mathfrak{S}'$.

Now, if K_0 is in *any* state $\mathfrak{S}^a(\bar{x}^a, U^a)$ in an ϵ-neighbourhood of $\mathfrak{S}'$ then it may be brought *reversibly* into a state $\mathfrak{S}^b(\bar{x}', U^b)$. Specifically, the deformation coordinates, being freely adjustable, are to be varied according to the prescription $x_k = x_k^a + \lambda(\bar{x}_k' - x_k^a)$, $(k = 1, 2, \ldots, n-1)$, $0 \leqslant \lambda \leqslant 1$. The work done by K_0 in this process goes to zero with ϵ since, by hypothesis, the forces P_k are finite and the range of variation of every deformation coordinate goes to zero with ϵ. Hence $U^b - U^a$, and therefore $U^b - U'$, goes to zero with ϵ. If ϵ be chosen sufficiently small U^b will certainly be greater than U''. Hence $\mathfrak{S}^b$ and therefore $\mathfrak{S}^a$ will in every case be accessible from $\mathfrak{S}'$. Recalling the arbitrariness of the choice of $\mathfrak{S}^a$ this result is however in conflict with (51.1). The assumption that a state of the kind $\mathfrak{S}''$ existed was therefore false. All states $\mathfrak{S}''$ on $\mathfrak{L}$ which have $U'' < U'$ are therefore inaccessible from $\mathfrak{S}'$.

(*b*) The line $\mathfrak{L}$ may be adopted as a line of reference. If $\mathfrak{S}''(\bar{x}'', U'')$ is now any arbitrarily prescribed state of K_0 it may be inferred directly that in any reversible transition from $\mathfrak{S}''$ to a state $\mathfrak{S}^*(\bar{x}', U^*)$ the value of U^* of U is in fact uniquely determined. Were this not the case one would have two distinct but mutually accessible states on $\mathfrak{L}$, in conflict with the conclusion reached earlier. To any state there thus corresponds a unique 'adjoint state' on $\mathfrak{L}$, i.e. in the sense in which $\mathfrak{S}^*$ is the state adjoint to $\mathfrak{S}''$.

Any state $\mathfrak{S}$ may now be labelled by the value s of the energy of its adjoint state. If two states $\mathfrak{S}'$, $\mathfrak{S}''$ have labels s', s'' respectively then $\mathfrak{S}'' \not< \mathfrak{S}'$ if and only if $s' < s''$. This follows at once from the validity of the corresponding relation for the adjoint states. Of course $\mathfrak{S}' = \mathfrak{S}''$ if and only if $s' = s''$. With the particular (though arbitrarily chosen) $\mathfrak{L}$ agreed on previously, s is a definite function $s(\bar{x}, U)$ of $\mathfrak{S}$. If $\mathfrak{S}'(\bar{x}', U')$ is some given state then, in accordance with what has just been said, only those states $\mathfrak{S}(\bar{x}, U)$ will be quasi-statically (and therefore reversibly) accessible from $\mathfrak{S}'$ for which

$$s(\bar{x}, U) = s(\bar{x}', U'). \tag{52.1}$$

On the other hand the $\bar{x}$ are freely variable, so that (52.1) is one member of an everywhere dense family of hyper-surfaces, one such surface passing through every point of $\mathfrak{L}$. Consequently, for every state $\mathfrak{S}''$ in an ϵ-neighbourhood of $\mathfrak{S}'$, $s''-s'$ tends to zero with ϵ, i.e. $s(\bar{x}, U)$ is a continuous function of the coordinates. It may be called the *empirical entropy function* of K. In place of s one might take any convenient monotonically increasing continuous function of s as the empirical entropy: doing so would be equivalent to choosing a different set of reference states in place of $\mathfrak{L}$ from the outset. Of course, one may eventually replace U by another non-deformation coordinate, and write $s(x)$ for the entropy function.

53. Principle of increase of empirical entropy

Suppose $\mathfrak{S}'$ is the initial state and $\mathfrak{S}''$ the final state of a transition of K_0. Of course one must have $\mathfrak{S}' < \mathfrak{S}''$, and so $s'' \geqslant s'$. Thus, *the final value of the entropy cannot be less than its initial value.* This is none other than the Entropy Principle of Section 44, though stated here with reference to the empirical entropy. It thus appears now almost trivially as a by-product of the arguments leading to the definition of $s(x)$.

54. The integrability of dQ

According to the results of Section 52*b* the representative curves of all quasi-static transitions from an arbitrary initial state $\mathfrak{S}'(x')$ must lie in the fixed hyper-surface (52.1) through $\mathfrak{S}'$. On the other hand such transitions satisfy the equation $dQ = 0$ of Section 30*a*. This means that $dQ = 0$ has $s(x) = \text{const.}$ as a single algebraic equivalent, i.e. there must exist a function $\lambda(x)$ such that identically

$$dQ = \lambda \, ds, \tag{54.1}$$

so that λ is an integrating denominator of dQ. This state of affairs was discussed in general terms in Section 33. However, one may simply argue as follows. Consider an infinitesimal element of *any* quasi-static transition. In general dQ and ds will both differ from zero. However, whenever $dQ = 0$ one has a reversible adiabatic transition between neighbouring states, so that $ds = 0$. In other words, whenever the otherwise arbitrary dx_k are such as to make

$dQ = 0$ then ds will also be zero. Write $ds = \Sigma \Xi_i dx_i$ and recall that $dQ = \Sigma X_i dx_i$. Choose $dx_k = 0$ $(k > 2)$, so that

$$X_1 dx_1 + X_2 dx_2 = 0$$

requires $\Xi_1 dx_1 + \Xi_2 dx_2 = 0$. Eliminating dx_1/dx_2 one therefore has to have $X_1/\Xi_1 = X_2/\Xi_2$. Alternatively taking other pairs of differentials into account in the same way it follows that X_i/Ξ_i has the same value for all i, $\lambda(x)$, say. Then

$$dQ = \Sigma X_i dx_i = \Sigma \lambda \Xi_i dx_i = \lambda ds,$$

which is just (54.1) again.

55. Consequences of integrability reviewed

(*a*) With (54.1) one has evidently arrived again at the stage reached in Chapter 5 at the end of Section 39. The reader may therefore simply turn back and take up the account of the consequences of the Second Law at Section 40. However, one is actually farther ahead here in as far as the Entropy Principle has, in substance, already been obtained. It may therefore not be out of place to go a little way beyond (54.1), by way of review.

First, then, let K_C be a compound standard system composed of standard systems K_A and K_B. Empirical entropy functions, s_A, s_B, s_C have been determined, and the coordinates of K_C may then be taken as $x_1, \ldots, x_{n-2}, y_1, \ldots, y_{m-2}, s_A, s_B, t$, where t is the empirical temperature of K_C, $t = t_A = t_B$. For each system separately one has a relation of the form (54.1), whence, recalling (24.2), one has identically

$$\lambda_C ds_C = \lambda_A ds_A + \lambda_B ds_B. \tag{55.1}$$

By inspection, s_C thus depends only on s_A and s_B, whilst the integrating denominators can then depend on other coordinates only in such a way that they will not in fact appear in (55.1). Any such coordinate must be common to all three systems, which leaves only t. Thus λ_A, λ_B and λ_C have a common factor which depends on t only, i.e.

$$\lambda_A = T(t)\phi_A(s_A), \quad \lambda_B = T(t)\phi_B(s_B),$$
$$\lambda_C = T(t)\phi_C(s_A, s_B). \tag{55.2}$$

Writing $S_A = \int \phi_A(s_A) ds_A$, etc., (54.1) becomes

$$dQ = TdS, \tag{55.3}$$

which applies in the first place to K_A and K_B. However one easily convinces oneself that s_C and ϕ_C depend on the sum of S_A and S_B only, so that (55.3) applies also to K_C. $T(t)$ and $S(s)$ are the *absolute temperature* and *metrical entropy* functions; and entropy is additive in the sense that

$$dS_C(= dQ_C/T) = dS_A + dS_B. \tag{55.4}$$

(*b*) As regards the Entropy Principle one may simply think of the isentropics $s(x) = \text{const.}$ to be labelled by the value of S rather than of s; and the *Entropy Principle for standard systems* obtains. The arbitrary constant a hidden in (55.3),

$$TdS = (aT)d(a^{-1}S),$$

has to be so chosen that the entropy actually increases in an irreversible transition. If one considers neighbouring states of the sequence $\mathfrak{L}$ of Section 52*a*, one certainly has

$$dS = S(\bar{x}', U+dU) - S(\bar{x}', U) > 0.$$

If the same states be linked reversibly (say by reversible thermal conduction) one has to have $dQ > 0$. It follows that

$$T > 0. \tag{55.5}$$

In this way the main results of Sections 40–45 inclusive have been recovered. They were there dealt with in more detail and reference may be made to them as required. However, Section 46–50 should now be read, if they have not been read before.

56. Existence of $s(x)$, given a weak substitute for the First Law

The part which the Second Law by itself plays in establishing the ordering of states under discussion may be brought out more clearly by abandoning the First Law, and so the existence of U. The point at issue is this: that the Second Law remains meaningful even if the First Law should happen to be false; and that an empirical entropy function and an Entropy Principle can therefore still be established, provided certain assumptions are introduced which will yield the element of continuity which was previously contained in the continuity of U and the boundedness of its derivatives.

The additional assumptions just referred to may be of various kinds. In this section the somewhat naïve procedure of simply tailoring them to the present requirements will be adopted. They will provide just enough information to make the arguments of Section 52 feasible again more or less as they stand. Yet they do not imply the First Law, and to that extent they are—in the present context—a weak substitute for it.

First of all, let it be recalled that, as before, only adiabatically isolated standard systems are being contemplated. Such a system K_0 has $n-1$ deformation coordinates $\bar{x}$, and one non-deformation coordinate ξ. The $\bar{x}$ are freely adjustable, both quasi-statically and otherwise. As regards ξ it will be supposed that it may be so chosen that the following assumption is valid:

$$\mathfrak{S}(\bar{x}, \xi) < \mathfrak{S}'(\bar{x}, \xi') \quad \textit{if} \quad \xi' \geqslant \xi. \tag{56.1}$$

This, then, simply replaces (51.2).

As in Section 52*a* one now considers a sequence of states $\mathfrak{S}''(\bar{x}', \xi'')$, i.e. states which lie on a line $\mathfrak{L}$ in R_n, through some state $\mathfrak{S}'(\bar{x}', \xi')$, and parallel to the ξ-axis. Proceeding as before one arrives at a stage of the argument in which one has to say something about the relationship between the values of ξ of two mutually accessible states, one of which lies anywhere in an ϵ-neighbourhood of $\mathfrak{S}'$ and the other on $\mathfrak{L}$. This can now be done only by a fiat: it is contained in the second assumption that

If $\mathfrak{S}'(\bar{x}', \xi')$, $\mathfrak{S}''(\bar{x}'', \xi'')$, $\mathfrak{S}^*(\bar{x}', \xi^*)$ *are such that* $d(\mathfrak{S}', \mathfrak{S}'') < \epsilon$, $\epsilon > 0$, *and* $\mathfrak{S}^* = \mathfrak{S}''$ *then* $\xi^* - \xi'$ *goes to zero with* ϵ. (56.2)

Hereafter everything goes much as before except in as far as ξ everywhere replaces U. In particular one has

$$s(\bar{x}, \xi) = s(\bar{x}', \xi') \tag{56.3}$$

in place of (52.1).

The end result of all this is that an ordering has been established for the states of K_0, reflected in the empirical entropy function

$s(\bar{x}, \xi)$ and its concomitant Entropy Principle: but the weaker, essentially qualitative, statements (56.1) and (56.2) have replaced the much stronger (quantitative) First Law and its ancillary assumptions. To this extent a more clear-cut separation between the First and Second Laws has been achieved. Of course, it does not seem that one can go any further without drawing upon the First Law, that is to say, as soon as *metrical* entropy and *absolute* temperature are to be defined.

57. On a strengthened version of the Second Law

Recall the remarks made at the beginning and end of the preceding section. The view there taken was this: that it is the principal function of the Second Law to ensure that the relation of adiabatic accessibility is not trivial; and that this non-triviality extends already to every neighbourhood of an arbitrarily chosen state $\mathfrak{S}$ in $\mathscr{S}$. It is worth enlarging a little on this point of view.

Let the symbols $<$ and $\not<$ used hitherto be replaced by the symbols $\prec$ and $\not\prec$ respectively. As pointed out in Section 37, the accessibility relation '$\prec$' is reflexive and transitive. It induces another relation '$\equiv$' [previously written '$=$'] the meaning of which is implied by the statement: $\mathfrak{S} \equiv \mathfrak{S}'$ if and only if $\mathfrak{S} \prec \mathfrak{S}'$ and $\mathfrak{S}' \prec \mathfrak{S}$. This latter relation is not only reflexive and transitive, but also symmetric; so that it is what is called an *equivalence relation* (in $\mathscr{S}$). In other words, it is true that (i) $\mathfrak{S} \equiv \mathfrak{S}$, (ii) $\mathfrak{S} \equiv \mathfrak{S}'$ *and* $\mathfrak{S}' \equiv \mathfrak{S}''$ *entails* $\mathfrak{S} \equiv \mathfrak{S}''$, and (iii) *if* $\mathfrak{S} \equiv \mathfrak{S}'$ *then* $\mathfrak{S}' \equiv \mathfrak{S}$; and one sees this also at once on recalling that '$\equiv$' is the relation of mutual (adiabatic) accessibility. Write $[\mathfrak{S}]$ for the set of all states $\mathfrak{S}'$ in $\mathscr{S}$ such that $\mathfrak{S}' \equiv \mathfrak{S}$, so that $[\mathfrak{S}]$ is an *equivalence class*. Then $\mathscr{S}$ splits up into equivalence classes $[\mathfrak{S}]$, $[\mathfrak{S}']$,.... These are mutually disjoint, that is to say, no two of them have any elements (i.e. states) in common. The set of all such classes $[\mathfrak{S}]$ will be denoted by Γ.

Now recall (37.1), according to which no mutually inaccessible pairs of states exist. Introduce an order relation '$<$' in Γ through the statement '$[\mathfrak{S}] < [\mathfrak{S}']$ *if and only if* $\mathfrak{S}' \not\prec \mathfrak{S}$'. Then $[\mathfrak{S}] < [\mathfrak{S}']$ and $[\mathfrak{S}'] < [\mathfrak{S}]$ cannot hold simultaneously, for otherwise no member of the first could be accessible from any member of the second and vice versa, contrary to (37.1). Similarly, if $[\mathfrak{S}]$ and $[\mathfrak{S}']$ are distinct then either $[\mathfrak{S}] < [\mathfrak{S}']$ or $[\mathfrak{S}'] < [\mathfrak{S}]$. Γ therefore is a linearly ordered set, or *chain*, with respect to the relation '$<$'.

All this is little more than a set of definitions. Where, one may well ask, does the Second Law make its appearance? In a certain sense it has in fact not yet entered the scene at all; and nothing prevents the relations introduced above from being nugatory. Indeed, they only achieve their purpose once the Second Law is adduced to assert that when one is concerned with any real physical system then more than one distinct equivalence class actually goes to make up Γ.

Of course, the Second Law contains more than this, in as far as it does not merely assert that, given any $\mathfrak{S}$, states inaccessible from $\mathfrak{S}$ exist, but rather that such states are to be found already in every neighbourhood of $\mathfrak{S}$. However, even this is not yet enough to lead to an empirical entropy function. As on previous occasions additional assumptions have to be made. They were concerned with aspects of continuity, i.e. they were essentially of a topological character. A suitable assumption in the present context is the following: if a state $\mathfrak{S}'$ is inaccessible from a state $\mathfrak{S}$ then every state in a certain neighbourhood of $\mathfrak{S}'$ is inaccessible from every state in a certain neighbourhood of $\mathfrak{S}$. This then endows the relation '$<$' with 'continuity'. Altogether one now has the Second Law and two ancillary assumptions:

For any $\mathfrak{S}$ *and* $\epsilon > 0$ *there exists* $\mathfrak{S}'$ *such that* $\mathfrak{S} \nprec \mathfrak{S}'$ *and* $d(\mathfrak{S}, \mathfrak{S}') < \epsilon$; (57.1)

If $\mathfrak{S} \nprec \mathfrak{S}'$ *then* $\mathfrak{S}' \prec \mathfrak{S}$; (57.2)

If $\mathfrak{S} \nprec \mathfrak{S}'$ *then there exists* $\delta > 0$ *such that* $d(\mathfrak{S}, \mathfrak{S}_1) < \delta$ *and* $d(\mathfrak{S}', \mathfrak{S}_1') < \delta$ *imply* $\mathfrak{S}_1 \nprec \mathfrak{S}_1'$. (57.3)

Equations (57.1–3) may be jointly regarded as a *strengthened version of the Second Law*, and it suffices to establish the existence of an empirical entropy function $s(\mathfrak{S})$ with the already familiar properties (i) $s(\mathfrak{S})$ is real-valued and continuous; (ii) $s(\mathfrak{S}) < s(\mathfrak{S}')$ if and only if $\mathfrak{S}' \nprec \mathfrak{S}$; (iii) there exists no $\mathfrak{S}$ and no $\delta > 0$ such that $s(\mathfrak{S}') \geqslant s(\mathfrak{S})$ for every $\mathfrak{S}'$ for which $d(\mathfrak{S}, \mathfrak{S}') < \delta$. The reader who is interested in seeing in detail how this may be done may consult the following paper: H. A. Buchdahl and W. Greve, *Zeitschrift für*

Physik, **168** (1962), 386. Evidently one is now dealing throughout with states considered as abstract objects, and the coordinates recede into the background; whilst the First or Zeroth Laws are of course never referred to at all. In this way the primary physical meaning of the entropy as a function reflecting the existence of a linear ordering amongst the classes of mutually accessible states of K_0 is most clearly brought out.

CHAPTER 7

THE THIRD LAW

58. On the behaviour of systems as $T \to 0$

(*a*) It is obvious that the zero of the absolute temperature presents a special problem both from a formal and from a physical point of view. Formally it must be regarded apart from all other values of T if for no other reason than that it represents a singularity of the integrating factor T^{-1} of dQ; so that, physically, one has to inquire into the behaviour of systems as T tends to zero in order to be able to say anything at all about the behaviour of the function S in this limit. In this connection it remains of course to be seen whether any statement considering the behaviour of a system *at* $T = 0$ is meaningful, for no such statement will be meaningful unless the system can in fact attain zero temperature. [The unqualified term 'temperature' will now generally be understood to refer to the absolute temperature.] Here a word might be said about the general notion of 'the attainment of the absolute zero'. Temperatures of the order of two micro-degrees have been reached in practice and it might be argued that temperatures of this order of magnitude are already so small that an assertion to the effect that $T = 0$ cannot be reached becomes effectively empty, especially as one might well be able to construct in the future some 'refrigerator' with the aid of which temperatures of the order of 10^{-7}, or 10^{-8},... degrees K might be reached. This argument rests upon the inherent assumption that temperatures as low as these are physically not distinct from $T = 0$. Clearly, however, a physical quantity can be said to be 'small' only in comparison with some standard; that is to say, its value is dependent upon the particular units employed in the process of measurement, quite apart from the possibility of making paper-and-pencil redefinitions at some later stage. The 'distance' d between two points may be 'large' or 'small' (numerically) according as the metre or the light year be taken as unit; and there is no reason why subsequently d^{-2} or $e^{1/d}$ for instance could not be taken as the 'distance'. If one of these functions were absolutely preferred for some reason then one might call it the 'absolute distance function', and all others 'empirical distance functions'. There is, however, no way of declaring a particular value of any one of these to be *a priori* small or large.

Phenomenologically the question is whether the properties of a system depend significantly upon T when T is reduced below a certain value: experiment shows that in this sense the lowest temperatures experimentally accessible at present are still 'large'. The onset of superconductivity and the specific heat anomalies of paramagnetic salts are characteristic examples of relevant phenomena. As a matter of interest, the result (50.8), for instance, suggests that in comparing absolute temperatures one should consider their ratio rather than the difference between them. Thus, one is familiar with the properties of substances at room temperature T^* ($=300$ °K, say), and one may consider temperatures differing from this by powers of 10. Leaving 'obvious' phenomena such as melting and evaporation aside, there is a characteristic rise of the specific heat c (at constant volume) of common gases such as H_2, CO, CO_2, N_2, which takes place at temperatures of the order $10T^*$. In the other direction, the specific heats c of the metallic elements fall below their 'classical values' as the temperature is decreased. Though there is a fairly wide scatter about the mean, c in fact decreases to one-tenth of its classical value at an average temperature of the order of $10^{-1}T^*$. Next, in the same vein, the average transition temperature of the superconducting elements is $10^{-2}T^*$. In the region between this temperature and $10^{-1}T^*$ one finds the λ-points of He^3-He^4 mixtures, whilst below this (i.e. down to $10^{-5}T^*$ and lower) there are pronounced maxima in the specific heats of certain magnetic substances. In short, it is a mistake to think of the difference between, say, 10^{-2} and 10^{-3} degrees as being phenomenally in any way 'comparable' with the difference between, say, 300·009 and 300·000 degrees; and it might be better deliberately to adopt an empirical temperature such as that defined by

$$t = T_1 \ln(e^{T/T_1}-1), \tag{58.1}$$

where T_1 is a disposable positive constant.

(*b*) The discontinuity of the integrating factor of dQ as $T\to 0$, and the universality of the absolute temperature function indicate that the behaviour of all substances capable of thermodynamic description must have certain common features, and the complex changes which are exhibited by many systems as T is reduced towards zero can hardly come as a surprise. A few of them were mentioned above, but the list is very far from exhaustive. Now it might be that these common features can be subsumed in some way under some general statement which is of universal application, yet avoids the complexities of a detailed enumeration of the properties of specific substances. Such a statement would naturally be regarded as a further Law of Thermodynamics, the Third Law. It would differ from the Laws preceding it in only one important respect, namely, it would not give rise to new physical quantities but would merely impose limitations upon those already defined.

59. The problem of the attainability of zero temperature

(*a*) As already remarked in the previous section it is meaningless to talk about the behaviour of a system K when $T = 0$ if K cannot be brought into any state of zero temperature. The question of whether $T = 0$ can or cannot be attained is therefore in itself of importance. To begin with it may be noted that $U(\bar{x}, T)$, regarded as a function of T, tends to its minimum value as $T \to 0$. It is not obvious that the limiting value of U is finite. That it is in fact so was, however, already assumed in Section 25. The minimum value of U depends of course on the values of the deformation coordinates $\bar{x}$.

Let K consist of a fixed amount of a certain substance, and choose the coordinates to be T and some deformation coordinate x, so that

$$TdS = dU+Pdx = U_{,T}dT+(P+U_{,x})dx. \tag{59.1}$$

Then if dT, dx are the coordinate differences of two states on a given isentropic

$$dT = -[(P+U_{,x})/U_{,T}]dx. \tag{59.2}$$

From (59.1), however, the integrability condition on dS is

$$P+U_{,x} = TP_{,T}, \tag{59.3}$$

so that

$$dT = -c^{-1}TP_{,T}dx. \tag{59.4}$$

The 'classical' (i.e. high temperature) behaviour of substances appears, if anything, to suggest the constancy of c. If c in fact tended to a non-zero value as $T \to 0$ then (59.4) would show that no adiabatic process could lead to $T = 0$, granted that $P_{,T}$ remains finite. One need not consider diathermic processes since a reduction of temperature towards zero by conduction would require some other system already at a lower temperature than that of the system of interest. Nor is an asymptotic approach towards zero to be contemplated, because no matter what the 'final' value of T may be, so long as it is not zero there is no proper criterion for saying that it is 'small enough', as explained at length in Section 58*a*. [Thus one can only assert that $T = 0$ either is or is not attainable. To qualify the assertion—as is often done—by adding the phrase 'in a finite number of operations' seems rather odd, to say the least; the more so as one can in any case never carry out an infinite number of operations.]

(*b*) Measurements carried out at lower and lower temperatures show that the specific heat c in fact tends to zero with T, and that it does so at least linearly. Equation (59.4) then shows that by a mere quasi-static change of the deformation coordinate the temperature of K_0 could be reduced to zero, *unless* $P_{,T}$ goes to zero. On the other hand, from (59.1) and (59.3), i.e. from the integrability condition on $d(TS-U) = SdT+Pdx$,

$$P_{,T} = S_{,x}. \tag{59.5}$$

Thus, a necessary condition for the unattainability of $T = 0$ is that the entropy should become independent of x, or, under more general circumstances, of all the deformation coordinates as $T \to 0$.

One thus sees that the impossibility of reducing T to zero is bound up with the behaviour of the metrical entropy as $T \to 0$. Moreover, if one calculates the entropy in this limit by imagining the temperature to be reduced by thermal conduction, the (formal) integral $\Delta S = \int T^{-1} c\, dT$ will not converge unless $c \to 0$ as $T \to 0$. (The finiteness of ΔS does not however require that c vanish at least as fast as T.)

60. The Third Law

(*a*) The stage has now been reached for a formulation of the Third Law which will reflect as far as possible the experimental situation under very general circumstances. To begin with, the vanishing of c will be assured if the entropy tends to a finite value as $T \to 0$. Next, various methods of producing lower and lower temperatures by means of adiabatic processes indicate that the properties of substances are just such as to prevent the absolute zero from being reached. The coefficient of expansion (at constant pressure) goes to zero with T, whence, granted that the isothermal compressibility does not vanish, the derivative $P_{,T}$ in (59.5) also goes to zero with T. Likewise, as regards a (homogeneous isotropic) paramagnetic substance whose magnetic moment is $\mathcal{M}$ in the direction of a uniform external magnetic field $\mathcal{H}$, take T as the non-deformation coordinate, whilst as deformation coordinates take $-\mathcal{M}$ and the volume V. The generalized force corresponding to the coordinate $-\mathcal{M}$ is $\mathcal{H}$ (see Section 83(viii)). Experiment shows that

$$\mathcal{M}_{,T} \to 0 \quad \text{as} \quad T \to 0, \tag{60.1}$$

and so the process of adiabatic demagnetization will not lead to the absolute zero. The kind of experiment here envisaged thus indicates that for every deformation coordinate x_k

$$S_{,x_k} \to 0 \quad \text{as} \quad T \to 0. \tag{60.2}$$

Hitherto, constitutive coordinates (implicitly referred to at the end of Section 3) have been left out of account. These will be considered in the next chapter in the context of the problem of chemical and phase-equilibrium. The study of chemical reactions, that is to say, of the conditions which obtain at equilibrium, indicates that as $T \to 0$ the entropy becomes independent also of the numbers n_k which specify the amounts of the various chemical constituents present (see Section 65), so long as the total amount of matter in the system remains constant. (Actually the validity of this conclusion is subject to a certain restriction which will be examined in Sections 61*b* and 94*c*.) Here it will suffice to illustrate the issue at stake by means of a simple example. It is desired to compare, as $T \to 0$, the sum of the entropies of a gram-atom of silver and a gram-atom of iodine with the entropy of a gram-molecule of silver iodide. To be able to do so one will have to be able to transform the given quantity of AgI *reversibly* into separate samples of silver and iodine. This can be done in principle. The AgI is slowly heated to a temperature at which it is, to all intents and purposes, completely dissociated. The gaseous silver and iodine are then separated by means of semipermeable membranes and afterwards cooled slowly to as low a temperature as feasible. The question then is: what is the magnitude of the entropy difference

$$\Delta S = S_{\text{AgI}} - S_{\text{Ag}} - S_{\text{I}}, \tag{60.3}$$

in the limit in which the initial and final temperatures go to zero? Since the entire process was reversible ΔS can be calculated by measuring all the specific heats, heats of melting, and heats of vaporization involved. However, the experiment just described is hardly feasible in practice. Yet the assumption that $\Delta S = 0$ brings with it testable consequences (cf. Section 94): and it is found to be satisfied. In certain analogous cases direct experiments can in fact be carried out and they lead to similar conclusions.

(*b*) Returning to the remarks made at the beginning of Section 59, any set of mutually isometric states contains a state such that the value of its energy is less than that of any other member of the set. Such a state shall be called 'a state of least energy'. Then it is not difficult to convince oneself that the various experimental results referred to above, including by implication even the fact that the energy has a finite minimum, can be summed up in the

> *Third Law*: *the entropy of any given system attains the same finite least value for every state of least energy.* (60.4)

This statement must be regarded as formal in the sense that it concerns itself with the properties of certain *functions* as $T \to 0$. One cannot make any sensible statements about transitions at the absolute zero, for no system can attain this temperature. To a certain degree it was just the recognition that this is the case which led to (60.4). By way of amplification, the latter states that as $T \to 0$ S goes to a finite minimum S_0, where S_0 is independent of all other coordinates. As remarked previously the most favourable process for reducing the temperature of a system below that of any other available system will be one which is adiabatic. However, since the entropy of all states with $T \neq 0$ is greater than S_0 the Entropy Principle at once implies that *it is impossible to reduce the absolute temperature of any system to zero.*

If one grants the vanishing of the specific heats as $T \to 0$, this unattainability statement can be regarded as a formulation of the Third Law. In that case one is led to the harmonious conclusion that the Third Law, like the First and the Second, is a Law of Impotence.

61. Further remarks concerning the Third Law

(*a*) The formulation (60.4) of the Third Law is such as to lead most directly, and unambiguously, to equations characterizing the observed behaviour of systems at very low temperatures. It is scarcely necessary to demonstrate this in detail, since to do so would require little more than to go again through what was said earlier in this chapter. Very briefly, the convergence of S to a finite limit S_0 as $T \to 0$ requires at any rate that c go to zero with T. That S_0 does not depend on the other coordinates entails the vanishing of quantities such as the isobaric coefficient of expansion; whilst with regard to chemical reactions it leads to the possibility of predicting the values of equilibrium constants (cf. Section 94) on the basis of calorimetric measurements alone.

If (60.4) is to be valid without blatant exceptions the equations (48.13) imply that the behaviour of the class of substances to which they relate certainly cannot be represented by (48.7) when T becomes sufficiently small. As a matter of fact the latter equations were at the outset stated to apply only when V and PV were sufficiently large. The whole problem will be considered afresh in Section 71.

(*b*) It will have been noticed that no reference whatever has been made to 'absolute entropy'. The reason for this is that such a quantity simply does not appear in the theory (and one can argue that this is equally true in statistical thermodynamics). Only entropy *differences* can be measured, and the value of the constant S_0 is not determinable. This being so, one might simply agree to the convention of setting $S_0 = 0$ in all cases. At first glance this step would seem to be quite unexceptionable, yet it entails an undesirable restriction upon the extent to which thermodynamic theory can be applied *in practice* to the description of physical systems. The italics are here intended to serve as a reminder that a pragmatic approach to the subject is essential, as has already been stressed repeatedly. Especially at low temperatures many substances have a tendency to exhibit metastability of one sort or another. A glass may retain its amorphous structure apparently indefinitely, though presumably it would ultimately become crystalline. Yet the time which may have to elapse before such a transformation will occur may be so large that it would be absurd not to regard the glass as a stable substance for the purposes of thermodynamic description, that is, in experiments in which the amorphous structure of the glass persists. More precisely, the glass may be regarded as being *in equilibrium* so long as experimental tests show that 'slow' transitions are in effect reversible. In particular, if the condition of the glass is described by the usual coordinates, the specific heat must be a definite function of them, irrespectively of whether the experiment is done today or tomorrow. In that case it is perfectly sensible to ascribe to the glass a definite entropy function, and the Third Law thereupon becomes applicable.

Now suppose—and this is possible in practice—that a certain substance M (glycerol, for example), upon being cooled below a certain temperature T_1, *either* remains in an amorphous condition *or* goes into a crystalline condition, depending, for the sake of argument, upon uncontrollable details of the experimental setup. Suppose further that both the crystalline form M^c and the amorphous form M^a are such that they may be regarded as being in equilibrium in the sense discussed above. Then, the specific heats c^c and c^a having been determined down to a sufficiently low temperature and extrapolated to $T = 0$ (validly, one hopes), the differences in the entropies

$$\Delta S^c = S_1^c - S_0^c, \quad \Delta S^a = S_1^a - S_0^a$$

can be calculated, the subscripts 1 and 0 referring to temperatures $T = T_1$ and 0 respectively. On the other hand, $S_1^c - S_1^a = \Lambda/T_1$, where Λ is the measured heat of transformation, the latter being supposed to occur reversibly at temperature T_1. Thus now

$$S_0^c - S_0^a = \Lambda/T_1 + \Delta S^a - \Delta S^c, \tag{61.1}$$

all quantities on the right being known. Actual experiment shows that in the kind of situation under discussion, namely one in which one is dealing with an inhibition due to very large viscosity,

$S_0^c - S_0^a$ is in fact not zero in general. Consequently, if one adopted the universal prescription $S_0 = 0$, then the properties of M^a would be excluded from thermodynamic description. Yet, below T_1 this exclusion is quite unnecessary, for S^a will behave in accordance with (60.4).

It should be borne in mind that the zero point entropies of different allotropic forms of a substance, both of which have a well-defined crystal structure, are equal. The kind of metastability which gives rise to the zero point entropy differences just considered is therefore of a special kind. It relates to situations in which large internal viscosities effectively prevent a pure substance from taking on a well-ordered physical structure. As already remarked, the Third Law in the form (60.4) may be applied to a substance lacking this specific structure, i.e. that of a pure crystal, under circumstances in which its actual structure, whatever this may be, remains generically fixed. In the context of chemical reactions this condition is implicitly not satisfied, and the Third Law can only be applied with due caution. This complication will be examined in Section 94*c*.

CHAPTER 8

POTENTIALS, CONSTITUTIVE COORDINATES, AND CONDITIONS OF EQUILIBRIUM

62. Thermodynamic potentials

(*a*) If one builds up the classical dynamical theory of systems in which the forces are conservative, starting from the equations of motion in the Newtonian form, the kinetic energy $\mathscr{T}$ and the potential $\mathscr{U}$ soon make their appearance. As the theory develops, certain other functions turn up quite naturally, in particular the Lagrangian $\mathscr{L}$ and the Hamiltonian $\mathscr{H}$. Note, however, that whereas $\mathscr{L}$ is a function of the coordinates and the velocities, $\mathscr{H}$ must be taken as a function of the coordinates and their conjugate momenta.

In thermodynamic theory one has a somewhat analogous situation (though in some respects the analogy leaves much to be desired). Thus, if K is some standard system with coordinates $x_1, x_2, \ldots, x_n$, $(x_n = T)$, the generalized forces $P_1, \ldots, P_{n-1}$, the energy U and the entropy S may be taken as known, having been determined empirically. Then it turns out to be convenient to define certain auxiliary functions, called *thermodynamic potentials*, of which the most prominent are

$$\text{(i) the } \textit{Helmholtz Function } F = U - TS, \tag{62.1}$$

$$\text{(ii) the } \textit{Gibbs Function } G = U - TS + \Sigma P_i x_i, \tag{62.2}$$

$$\text{(iii) the } \textit{enthalpy } H = U + \Sigma P_i x_i, \tag{62.3}$$

the summations extending from 1 to $n-1$. It will be seen that each of them has the dimensions of energy.

Recall that

$$T\,dS = dU + \Sigma P_i\,dx_i. \tag{62.4}$$

The total differentials of F, G and H therefore are at once

$$dF = -S\,dT - \Sigma P_i\,dx_i, \tag{62.5}$$

$$dG = -S\,dT + \Sigma x_i\,dP_i, \tag{62.6}$$

$$dH = T\,dS + \Sigma x_i\,dP_i. \tag{62.7}$$

In the first place one could now write down a large number of integrability conditions, but this may be left till later (Sections 81, 83 (viii)). At any rate, one has from (62.5)

$$S = -F_{,T}, \quad P_i = -F_{,x_i}. \tag{62.8}$$

Using the first of these, one has from (62.1)

$$U = F - TF_{,T}, \tag{62.9}$$

the so-called Gibbs–Helmholtz equation. The function $F(x)$ evidently characterizes K completely, for, given F, the functions P_i, S, U are derivable from it. It may therefore be properly referred to as a *characteristic function* of K; and here one might think of the analogy with Hamilton's point characteristic in geometrical optics. When one comes to G one assumes (cf. Section 12*a*) that the $n-1$ equations

$$P_i = \phi_i(x_1, x_2, \ldots, x_n) \quad (i = 1, \ldots, n-1)$$

can be solved uniquely for $x_1, \ldots, x_{n-1}$. Then, G, U, S may be written down as functions of $P_1, \ldots, P_{n-1}$, T. This having been done (62.6) gives

$$S = -G_{,T}, \quad x_i = G_{,P_i}, \tag{62.10}$$

and

$$U = G - TG_{,T} - \Sigma P_i G_{,P_i}. \tag{62.11}$$

Thus $G(P_1, \ldots, P_{n-1}, T)$ is also a characteristic function. Evidently in going over from F to G one does not merely have to add a term $\Sigma P_i x_i$ to F, but one has at the same time to make a change of independent variables. This is again reminiscent of the situation in optics when one goes over from the point to the angle characteristic.

One easily convinces oneself that $H(P_1, \ldots, P_{n-1}, S)$ is a characteristic function, but it is hardly worth while writing down the equations corresponding to (62.8, 9). Depending on the appropriate choice of independent variables the list of characteristic functions may be further extended. Moreover, it is sometimes convenient to consider *modified thermodynamic potentials*. First, however, let the following general notation be agreed upon: if X is any quantity, then

$$\tilde{X} \equiv X/T. \tag{62.12}$$

Then one has, for instance, the modified Helmholtz function

$$\tilde{F} = \tilde{U} - S, \tag{62.13}$$

so that

$$d\tilde{F} = -(\tilde{U}/T)\,dT - \Sigma \tilde{P}_i\,dx_i. \tag{62.14}$$

From this

$$\tilde{U} = -T\tilde{F}_{,T} \tag{62.15}$$

follows at once, but this is merely (62.9) again. $\tilde{G}$ may be dealt with similarly.

As a simple example of the way in which potentials turn up naturally, consider an isothermal transition from a state $\mathfrak{S}'$ to a state $\mathfrak{S}''$. If it is quasi-static one has from (62.5) that $dW = -dF$, since $dT = 0$, so that

$$W = -\Delta F. \tag{62.16}$$

In an adiabatic transition $W = -\Delta U$; but this is always true, whereas (62.16) is false when the transition is irreversible. In fact, generally, if T has the meaning assigned to it in (50.6, 12) and is constant,

$$W = -\Delta U + \int dQ = -\Delta U + T\int dQ/T \leqslant -\Delta U + T\Delta S,$$

by (50.6, 12), i.e.

$$W \leqslant -\Delta F. \tag{62.17}$$

Because of these results F is sometimes called the 'Helmholtz free energy'.

(*b*) It has already been remarked that every *characteristic* function is in the nature of things a function of a particular set of variables (see also Section 66*b*). It may be worth enlarging on this point a little. By way of example, consider the Gibbs Function $G(P, T)$ of a simple fluid of volume V. Thus from (62.10)

$$V = G_{,P}, \quad S = -G_{,T}. \tag{62.18}$$

G being given, its derivatives are known functions of P and T, so that the first of (62.18) is the equation of state of the fluid, whilst the second gives the entropy function explicitly as a function of P and T. Of course

$$U = G - TG_{,T} - PG_{,P}, \tag{62.19}$$

and so everything about the fluid is known. Now there is nothing to prevent one from prescribing G as a function of V and T, say

$G^*(V, T)$; the asterisk being intended to emphasize that the 'incorrect' variables are being used. The point is now that G^* is no longer a *characteristic* function. $G^*(V, T)$ being given, its derivatives $G^*_{,V} = A(V, T)$ and $G^*_{,T} = B(V, T)$ are known functions. P is now a function of V and T, and

$$VP_{,V} = A, \quad S = VP_{,T} - B.$$

These, however, determine P only to within an arbitrary additive function $\phi(T)$ of T alone, and then S and U are known only to within additive terms of the form $V\phi_{,T}$, $V(T\phi_{,T} - \phi)$ respectively.

Hitherto $n-1$ of the coordinates of a standard system were always supposed to be deformation coordinates. Now, as far as the definition of standard systems is concerned the central necessity is that it should be *possible* to choose a set of coordinates of which $n-1$ have deformative character. This does not mean that one *has* to choose them in this way. Consequently the term 'coordinates' should be understood now to refer to any admissible set of independent variables which can be introduced in place of the $x_1, \ldots, x_n$. By implication, when a particular characteristic function is being contemplated, a state of K will be a set of values of those coordinates which are appropriate to the characteristic function in question. It should be noted, however, that the introduction of coordinates other than the original $x_1, \ldots, x_n$ may bring about discontinuities which did not have to be dealt with before; e.g. continuity of $S(x_1, \ldots, x_{n-1}, T)$, $(x_k = \text{const.}, k \neq n)$ does not necessarily imply the continuity of $S(P_1, \ldots, P_{n-1}, T)$, $(P_k = \text{const.})$. One will still have piece-wise continuity, but the discontinuities will have to be allowed for separately (cf. Section 83 (vii)).

(*c*) Again by analogy with optical theory, the term 'conjugate variables' is sometimes convenient. Thus given a characteristic function denoted by the general symbol Ξ, let $\xi_1, \ldots, \xi_n$ be the appropriate coordinates. Then one may speak of the n quantities π_k defined by

$$\pi_k = \partial\Xi/\partial\xi_k \tag{62.20}$$

as the variables *conjugate* to ξ_k. For example, if Ξ happens to be G, then $(\pi_1, \ldots, \pi_{n-1}, \pi_n) \equiv (x_1, \ldots, x_{n-1}, -S)$.

63. The system as a black box

One may be unable in principle to 'look inside' a physical system of some sort in order to discover just why, in its interactions with its surroundings, it behaves in some particular fashion. Sometimes one may simply pretend not to be interested in finding out what goes on inside. In either case one often picturesquely describes the system as a 'black box'. For example, a beam of particles is aimed at a target and the scattered particles are observed at points remote from the target. Then one may seek to establish general relationships between the outgoing and ingoing beams without attempting a description of the detailed history of the particles as they pass through the scattering region: then the latter is a typical black box.

Now all the thermodynamic systems K considered hitherto have been treated as black boxes. It is true that in requiring the absence of internal adiabatic partitions one seems to be saying something about the interior of K. Still, this objection is largely verbal, for the condition in question might be replaced by the requirement that if a system K^* is in equilibrium with K when placed in diathermic contact with K then this equilibrium shall persist when K^* is brought into contact with any other part of K. In short, attention has been confined to the establishment of general limitations governing the interactions of a given system with its surroundings. That the system is 'given' is intended to imply that its generic structure is fixed, in particular that it is *closed*: meaning that it cannot exchange matter with its surroundings. At first sight it might be thought that the Entropy Principle is at variance with the present point of view in as far as it appears to say something about the system itself. It should however be kept in mind that the irreversibility of a transition is of interest largely because the restoration of the initial state implies that the final condition of the *surroundings* must differ from their initial condition.

Suppose next that a simple system K^* whose coordinates are T and P be experimentally investigated. For the sake of argument, let the pressure P be kept fixed at some value P_1. One then has the well-defined problem of determining the increase of volume V consequent upon an increase in T. Under certain circumstances one may find that at some temperature T_1 (depending upon P_1) the volume changes discontinuously. The corresponding 'specific heat at constant pressure' c_P, that is,

$$(dQ/dT)_{P=\text{const.}} = c_P = U_{,T} + PV_{,T}, \tag{63.1}$$

is then also discontinuous at $T = T_1$.

So long as K^* is regarded as a black box this state of affairs just has to be accepted. One will know how the system was originally constituted, in the sense that certain given amounts of substances were introduced into the enclosure of volume V. (In the case under consideration one might think of K^* as having been prepared by introducing a certain quantity of, say, water into an otherwise empty enclosure). The subsequent behaviour of K^*, as revealed by experiment in the manner just described, is however not further analysed; in particular the discontinuity of c_P is not traced back to anything that may be going on (at a macroscopic level) within the enclosure. This situation is, indeed, typical of the 'black box view'. The deformation coordinates and the one non-deformation coordinate are, as usual, under one's control, and a characteristic function comprises one's total knowledge of the system: just what goes on within it is ignored.

64. The system not as a black box

Whether or not the theory so far developed requires basic extension, the time has come to go beyond the attitude maintained hitherto, and to 'look inside' the system of interest. Take again the simple system K^* of the previous section. Upon examining the interior (supposed to contain water only) let it be found that liquid water but no ice is present. Then upon raising the temperature of K^* it will be observed that just when it reaches the value T_1 it cannot be raised any further until all the water has vaporized. A peculiarity in the behaviour of K^* has thus been traced to a feature of its constitution, viz. the presence of two distinct forms of water and the complete transformation of one into the other at a definite temperature (the pressure being given).

More generally a system may consist of any number of physically homogeneous parts, each of which has a definite physical boundary.

Any such part is called a *phase*. (Save under the most exceptional circumstances there can be only one gaseous phase.) The term 'boundary' is here to be understood in the geometrical sense, that is to say, it is the geometrical surface over which different phases are visibly in contact. One is therefore now admitting the possibility of the passage of matter across such a boundary: in the case of the standard system of Section 10*c*, on the other hand, no matter could be transferred from a phase contained in one enclosure to a phase contained in any other, since the boundary between them was material and impermeable. A phase, temporarily regarded as a subsystem, is 'open', i.e. the amount of matter in it is not fixed, as distinct from the situation which obtains with regard to proper subsystems. Once again returning to the example of K^*, even though the total amount of water was fixed in that case, that in the liquid phase was not. However, it is obviously of interest to know the amount of (liquid) water present when P and T are prescribed.

Once the openness of phases is taken into account, one may as well consider systems which are open as a whole. Moreover, upon concerning oneself with the internal constitution of systems, one has to allow for the possibility of the occurrence of *chemical reactions*. Any chemically distinct substance within a system is called a *constituent* of the system. In general, when a certain amount of a constituent is introduced into the system, some of it may lose its identity, owing to the possibility of molecular association or dissociation taking place. However, once the system is in a state (the usual coordinates having prescribed values) it is obviously of great practical importance to know just how much of each constituent is present in the system.

65. Constitutive coordinates. External and internal states

In view of what has just been said, there now arises the problem of making a convenient choice of the new variables intended to describe the internal constitution of systems. For the time being it suffices to consider a single phase (in equilibrium). This will generally consist of a mixture of chemical substances. Its composition could be described by giving the masses of the various constituents present in it. However, recall that a chemical reaction between compounds obeying the law of definite proportions (which alone will be considered) can always be represented by an equation of the form

$$\sum_i \nu_i C_i = 0, \tag{65.1}$$

where C_i is a generic symbol for the ith constituent, whilst the 'stoichiometric coefficient' ν_i is an integer, positive or negative according as C_i is regarded as appearing or disappearing in the reaction. Thus in the formation of water from oxygen and hydrogen, $C_1 \equiv H_2$, $C_2 \equiv O_2$, $C_3 \equiv H_2O$; $\nu_1 = -2$, $\nu_2 = -1$, $\nu_3 = +2$. It is evidently more convenient to associate with every constituent C_i its own unit of mass, namely a mass in grams whose value is numerically equal to the molecular weight of C_i. This unit is called a *mole*. Then the constitution of the phase is specified by prescribing the number of moles n_i of each constituent C_i present in the phase. Evidently when in a reaction n_i changes by Δn_i one has simply

$$\Delta n_1/\nu_1 = \Delta n_2/\nu_2 = \ldots \equiv \Delta\xi, \tag{65.2}$$

i.e. $\Delta n_i/\nu_i$ has the same value, $\Delta\xi$, say, for all i.

The required variables, describing the internal constitution of a phase, will be taken to be the numbers n_i (also called 'mole numbers'). They will be referred to either as *constitutive* or as *internal* coordinates. The coordinates describing the system *regarded as a black box*, e.g. $x_1, \ldots, x_{n-1}, T$, will, when necessary, be distinguished by the qualification 'external'. If a system contains several phases, then the mole numbers for each must be taken as coordinates, i.e. n_i^k is the mole number of the ith constituent in the kth phase. Thus a complete set of coordinates of a system will be $x_1, \ldots, x_n$; $n_1^1, \ldots, n_z^1$; $\ldots$; $n_1^p, \ldots, n_z^p$, there being p phases, each with z constituents; so that there are now

$$N = n + pz, \tag{65.3}$$

coordinates in all.

A *state* $\mathfrak{S}$ of the system is now a set of values of its N coordinates, equilibrium being implied as usual. However, it is sometimes useful to speak of the *external state* $\mathfrak{S}^e$ of K, which is a set of values of the external coordinates, and the *internal state* $\mathfrak{S}^i$, which is a set of values of the internal coordinates. $\mathfrak{S}^e$ therefore is what was previously denoted simply by $\mathfrak{S}$.

It is important to realize that the constitutive coordinates, unlike the external coordinates, are not, in general, freely variable. One sees this immediately upon considering a system which is closed as a whole. As the external state is varied, the redistribution of matter amongst the various phases will take care of itself, whether by chemical reaction or passage from one phase to another, so that under these conditions the internal state is already determined by the external state. On the other hand, up to a certain stage the internal coordinates of a given phase may be considered as independent and freely adjustable, in a sense implicit in the next section.

66. A single inert phase as open system. Chemical potentials

(*a*) As in the previous section a system consisting of only a single phase will be considered at first, so that the superscripts of internal coordinates may be suppressed. Since one is now mainly concerned with problems of physico-chemical equilibrium, the most useful choice of external coordinates is the pressure P of the phase, and its temperature T, since equilibria are commonly studied under conditions of given P and T. Now, the possibility of there being further external coordinates cannot be excluded. The phase might for instance contain magnetic substances, so that in the presence of an external magnetic field another coordinate has to be admitted, namely the magnetic moment $\mathscr{M}$ of the phase in the direction of the external field. However, in order not to encumber the various equations unnecessarily, they will be written as if there were only the single deformation coordinate V. In the more general case one need only replace PdV by $\Sigma P_i dx_i$ and make analogous modifications elsewhere in order to get completely general results.

To maintain reasonable coherence with the methods of earlier chapters, especially Chapter 5, one may proceed as follows. The phase under consideration is, in the first instance, contained at temperature T and volume V in an enclosure impermeable to matter. An (infinitesimal) amount dn_i of the ith constituent C_i is now to be introduced *reversibly* into the phase from outside the system K. To this end, imagine a tube to project from the enclosure, connecting the interior of K with that of an auxiliary enclosure K_i, containing only the substance C_i. The tube incorporates a semipermeable membrane which permits the passage of C_i alone. (Such membranes are realizable in practice, at any rate to the extent that perfect membranes of this kind may be admitted into the argument. The situation is somewhat analogous to that which arises in the context of the admissibility of enclosures which are strictly adiabatic.) The temperature of K_i, like that of K, is T, whilst the pressure is just such that C_i will not pass through the membrane. By decreasing the volume of K_i at an infinitesimal rate, with T and V held fixed, an amount dn_i of C_i will pass into K. The work done on K_i may now be looked upon as work done on K in the process of introducing the amount dn_i of C_i into it. Let this work be written as $\mu_i dn_i$.

Several constituents may be added to K simultaneously provided one has a separate auxiliary enclosure for each, together with the appropriate semipermeable membrane. Finally, if all the tubes are blocked and V changed alone, the work done on K is $-PdV$. Hence when amounts $dn_1, dn_2, \ldots$ of $C_1, C_2, \ldots$ are introduced into K, whilst the volume of the phase changes by dV, the work done by K will be

$$dW = PdV - \sum_{i=1}^{z} \mu_i dn_i. \tag{66.1}$$

The coefficient μ_i of dn_i is called the *chemical potential* of C_i, and it depends, in general, on P and T, *and* the constitution of the phase.

(*b*) For the time being, let it be supposed that no chemical reactions occur within K. Then when an amount dn_i has passed into the phase from outside then the amount of C_i actually present within it will have increased just by dn_i. In this case the internal coordinates reveal themselves as formally similar to the deformation coordinates: the internal 'configuration' of K is fully described by coordinates n_i which are freely variable, whilst dW is again a linear differential form, i.e. (66.1). The analogues of the generalized forces are just the chemical potentials with signs reversed.

Now let a quasi-static transition take place *adiabatically*. Then in virtue of the First Law, applied to the general situation now under investigation, the energy is defined, as before, through

$$dU = -dW. \tag{66.2}$$

Then the Second Law eventually yields

$$TdS = dU + dW. \quad (66.3)$$

Now recall the *definition* (62.2) of the Gibbs function

$$G = U - TS + PV. \quad (66.4)$$

This is, of course, retained unaltered. The total differential of this becomes immediately

$$dG = VdP - SdT + \Sigma\mu_i dn_i. \quad (66.5)$$

G is now a function of the coordinates $P, T, n_1, \ldots, n_z$. It is a characteristic function since, as before, all relevant thermodynamic functions pertaining to K are derivable from it by differentiation:

$$G_{,P} = V, \quad G_{,T} = -S, \quad G_{,n_i} = \mu_i, \quad (66.6)$$

$$U = G - TG_{,T} - PG_{,P}. \quad (66.7)$$

Now, however, the characteristic function also concerns itself with the internal constitution of the system. One may also consider other characteristic functions again, e.g. $F(V, T, n_1, \ldots, n_z)$, $U(V, S, n_1, \ldots, n_z)$, and so on; see, for example, Section 66*d*.

(*c*) Imagine $k-1$ identical copies of K to be placed in diathermic contact with K so that one has a compound system K^*. The total volume, entropy, and energy of K^* are evidently k times those of K. The removal of all internal partitions from K^* will have no effect since the pressure and temperature were, by construction, uniform throughout K^* to start with. (The assumption that surface effects are negligible must be satisfied here, but it should be recalled that it was already involved in the argument leading to the additivity of energy.) If n_i were the values of the internal coordinates of K then those of K^* will be kn_i. One arrives thus at the conclusion that if the values of the internal coordinates of two systems are in the ratio $k:1$, whilst the pressure and temperature are, respectively, the same for both, then the volume, energy and entropy, and therefore the Gibbs function have values which stand in the same ratio. This result will be valid whether k is integral or not.

Now take $k = 1 + \eta$, where η is infinitesimal. Then the internal coordinates of the two systems differ by $dn_i = \eta n_i$, $(i = 1, \ldots, z)$.

On the other hand the second system may equally well be thought of as having arisen from the first by the addition of amounts dn_i of C_i $(i = 1, \ldots, z)$ to it at constant pressure and temperature. The resulting change of G being ηG, one has from (66.5)

$$G = \sum_{i=1}^{z} \mu_i n_i. \tag{66.8}$$

If this be differentiated with respect to n_k one has

$$G_{,n_k} = \mu_k + \Sigma n_i \mu_{i,n_k}.$$

In view of the last member of (66.6) this becomes

$$\Sigma n_i \mu_{k,n_i} = 0. \tag{66.9}$$

This may be regarded as a differential equation for μ_k. To solve it, introduce quantities

$$c_i = n_i / \bar{n} \quad (i = 1, \ldots, z), \tag{66.10}$$

$$\bar{n} = \sum_{j=1}^{z} n_j. \tag{66.11}$$

One calls c_i the *concentration*, expressed as a mole fraction, of C_i. These concentrations are not independent of each other, since, identically,

$$\sum_{i=1}^{z} c_i = 1. \tag{66.12}$$

Choose $c_1, \ldots, c_{z-1}, \bar{n}$ as new independent variables in the equation (66.9). Then the latter reduces to

$$\mu_{k,\bar{n}} = 0. \tag{66.13}$$

In other words, regarded as a function of the internal coordinates, μ_k $(k = 1, \ldots, z)$ depends only on their ratio.

(*d*) At this stage brief mention may be made of the characteristic function

$$\Omega = \sum_{i=1}^{n-1} P_i x_i, \tag{66.14}$$

which is to be taken as a function of the deformation coordinates, the chemical potentials, and the absolute temperature. This function does not occupy a prominent place in the phenomenological theory. On the other hand it appears naturally in the statistical theory, namely, in the context of the grand canonical ensemble.

From (66.4) and (66.8)

$$\Omega = \Sigma \mu_i n_i - U + TS,$$

whence, using (66.1, 3),

$$d\Omega = \Sigma P_i dx_i + S\,dT + \Sigma n_i d\mu_i.$$

Therefore

$$P_i = \Omega_{,\,x_i}, \quad S = \Omega_{,\,T}, \quad n_i = \Omega_{,\,\mu i}. \tag{66.15}$$

Using (66.14) on the right of the first of these equations one gets the set of indentities

$$\sum_k x_k P_{i,\,x_k} = 0 \quad (i = 1, \ldots, n-1). \tag{66.16}$$

Every P_i is therefore a homogeneous function of degree zero of the deformation coordinates.

67. The active heterogeneous system

(*a*) The Gibbs function of a phase will be a definite function of $P, T, n_1, \ldots, n_z$, whether chemical reactions can occur within it or not. One should note in this context that when the phase is in a fixed state no reactions proceed within it: phenomenologically it is not meaningful to regard this situation as one in which reactions occur in opposite directions at equal rates. Consider a quasi-static transition between neighbouring states:

$$dG = G_{,P} dP + G_{,T} dT + \Sigma G_{,n_i} dn_i. \tag{67.1}$$

Here the dn_i are in general no longer the amounts of substances $d\mathfrak{n}_i$ passing across the boundary of the phase. The difference between dn_i and $d\mathfrak{n}_i$ will be the amount $\bar{d}n_i$ of C_i appearing within the phase as a consequence of whatever chemical reactions occur within it:

$$dn_i = d\mathfrak{n}_i + \bar{d}n_i. \tag{67.2}$$

From Section 66

$$dG = V\,dP - S\,dT + \Sigma \mu_i d\mathfrak{n}_i. \tag{67.3}$$

Evidently, the internal coordinates n_i are now strictly speaking not analogous to the external (deformation) coordinates. Their proper analogues are rather the amounts $-\mathfrak{n}_i$ of substances within the auxiliary enclosures K_i introduced previously.

If one chooses all the $d\mathfrak{n}_i$ to be zero, so that the phase is closed, (67.1, 3) give, for arbitrary dP and dT,

$$dG = G_{,P}dP + G_{,T}dT + \Sigma G_{,n_i}\bar{d}n_i = VdP - SdT, \quad (67.4)$$

so that $$G_{,P} = V, \quad G_{,T} = -S, \quad \Sigma G_{,n_i}\bar{d}n_i = 0. \quad (67.5)$$

On the other hand, inserting these now in (67.1) and comparing the result with (67.3), the arbitrariness of the $d\mathfrak{n}_i$ entails

$$G_{,n_i} = \mu_i, \quad (67.6)$$

so that (67.1) now reads

$$dG = VdP - SdT + \Sigma\mu_i dn_i, \quad (67.7)$$

and (66.5) continues to apply.

(*b*) Subject to the usual assumptions concerning the absence of surface effects and distance forces—the former will always be negligible if every phase is sufficiently large—the Gibbs function G of a heterogeneous system may now be taken to be the sum of the Gibbs functions G^k of the separate phases. To see this, one may imagine the phases to be temporarily separated from one another by the insertion of material boundaries. The usual addition theorems then yield the stated conclusion. Thus

$$G(P, T, n_1^1, \ldots, n_z^1, \ldots, n_1^p, \ldots, n_z^p) = \sum_{k=1}^{p} G^k(P, T, n_1^k, \ldots, n_z^k). \quad (67.8)$$

Formally all constituents are allowed to be present in all phases. If the situation is otherwise in a particular case, then the n_i^k in question are to be set equal to zero. One now has

$$dG = VdP - SdT + \sum_{k=1}^{p}\sum_{i=1}^{z} \mu_i^k dn_i^k. \quad (67.9)$$

The results of Section 66*c* continue to apply.

68. Conditions for physico-chemical equilibrium: unnatural states not admitted

(*a*) It is to be taken for granted that when a closed system K is in equilibrium it has a definite composition. In other words, when K undergoes a quasi-static transition in which the external coordinates vary in any prescribed manner, the internal coordinates automatically vary in a definite way. This much was already said in slightly different words at the end of Section 65. Now (67.9) holds for all quasi-static transitions. Under the present conditions K may, however, also be treated as a black box, in which case its Gibbs function makes no reference to internal coordinates, so that

$$dG = VdP - SdT. \tag{68.1}$$

In (67.9) the dn_i^k are now definite (linear) functions of dP and dT. Consistency of (68.1) with (67.9) requires that

$$\sum_{k=1}^{p} \sum_{i=1}^{z} \mu_i^k \, dn_i^k = 0. \tag{68.2}$$

It is worth emphasizing again that the dn_i^k here refer to differences in the values of the n_i^k corresponding to different *states* (of a closed system), i.e. they are the changes in the n_i^k brought about by the reversible adjustment of the external coordinates. Note also that if in the present context the characteristic function F, for example, had been used in place of G one would have arrived at (68.2) just the same.

Evidently (68.2) is a *necessary* condition for the existence of equilibrium. Whether it is also sufficient, when taken together with all relevant conditions, remains to be seen. Accordingly, consider first a single, chemically active phase. Equation (68.2) gives

$$\sum_{i=1}^{z} \mu_i dn_i = 0, \tag{68.3}$$

superscripts having been suppressed. (This is of course merely the third of (67.5) again.) If only a single reaction is possible, one has from (65.2)

$$dn_i = \nu_i d\xi, \tag{68.4}$$

whence (68.3) gives

$$\Sigma \mu_i \nu_i = 0. \tag{68.5}$$

On the other hand, the system will have been prepared initially by introducing amounts n_{i0} of C_i ($i = 1, \ldots, z$) into the enclosure. Thereafter the n_i have definite values

$$n_i = n_{i0} + \nu_i \xi, \tag{68.6}$$

the validity of this equation having nothing to do with the question of equilibrium. The equilibrium value of ξ then follows from the equation obtained by inserting (68.6) in (68.5), i.e.

$$\Sigma \nu_i \mu_i(P, T, n_{10}+\nu_1\xi, \ldots, n_{z0}+\nu_z\xi) = 0, \tag{68.7}$$

P and T having been prescribed. In this simple case, then, (68.2) turns out to be sufficient.

An analogous situation prevails with regard to the equilibrium between two phases of a single constituent. In that case (68.2) reads

$$\mu^1 \, dn^1 + \mu^2 \, dn^2 = 0, \tag{68.8}$$

subscripts having been suppressed. One also has

$$dn^1 + dn^2 = 0. \tag{68.9}$$

Provided that the component will pass at all between the two phases as P and T vary, (68.8, 9) are consistent only if

$$\mu^1 = \mu^2. \tag{68.10}$$

(*b*) At first sight it seems, however, that under more general circumstances the condition (68.2) is too weak. This may be seen most easily when several chemical reactions can occur between the components of a single phase. One then has simultaneously several equations of the form (65.1), i.e.

$$\sum_i \nu_i^{(r)} C_i = 0 \quad (r = 1, \ldots, R), \tag{68.11}$$

there being R independent reactions (see also the beginning of Section 88). Some of the $\nu_i^{(r)}$ may of course be zero. If $\Delta n_1, \Delta n_2, \ldots$ are the changes which occur as a consequence of the rth reaction, then

$$\Delta n_1/\nu_1^{(r)} = \Delta n_2/\nu_2^{(r)} = \ldots = \Delta\xi^{(r)} \tag{68.12}$$

is the equation corresponding to (65.2). In place of (68.6),

$$n_i = n_{i0} + \sum_r \nu_i^{(r)} \xi^{(r)}. \tag{68.13}$$

To (68.7) there then corresponds

$$\sum_r \sum_i \mu_i(P, T, n_1, \ldots, n_z) \nu_i^{(r)} \, d\xi^{(r)} = 0. \tag{68.14}$$

As it stands (68.14) is not sufficient to determine the equilibrium values of the $\xi^{(r)}$. However, it may be reflected that one may admit not only changes of P and T, but also addition of matter to the phase from without, provided one then writes, in the notation of (67.2),

$$\sum_{i=1}^{z} \mu_i \bar{d}n_i = 0 \tag{68.15}$$

in place of (68.3). This yields (68.14) as before. The $d\xi^{(r)}$ depend linearly on dP, dT, $d\mathfrak{n}_1, \ldots, d\mathfrak{n}_z$, all of which may be chosen arbitrarily. *If* one now makes a small concession in the form of the assumption that the $d\xi^{(r)}$ can be given arbitrary values independently of each other (i.e. by suitably choosing dP, dT, $d\mathfrak{n}_1, \ldots$), then (68.14) gives R separate conditions

$$\sum_{i=1}^{z} \mu_i \nu_i^{(r)} = 0 \quad (r = 1, 2, \ldots, R). \tag{68.16}$$

From these the equilibrium values of the $\xi^{(r)}$ may be determined.

What has emerged is that equilibrium values of the concentrations can be determined, provided one grants that the different chemical reactions are mutually independent in the sense that the *degrees of advancement* $\xi^{(r)}$ can be changed independently of each other by arbitrary amounts $d\xi^{(r)}$ by suitably choosing

$$dP, dT, d\mathfrak{n}_1, \ldots, d\mathfrak{n}_z.$$

The general heterogeneous system may be dealt with along similar lines. Subject to the assumption made above it would seem therefore that (68.2) is not only necessary but also sufficient. However, one hiatus remains: an apparently meaningful solution of (68.2) may have only formal significance, in that it may represent a physically unrealizable state. It is then analogous to the static equilibrium of a mechanical system which corresponds to a maximum of its energy function. This also is physically unrealizable, granted that external disturbances can never be completely excluded. Accordingly the whole problem of physico-chemical equilibrium will be treated afresh in the next section from an alternative point of view.

69. Conditions for physico-chemical equilibrium: unnatural states admitted

(*a*) In the preceding section equilibrium conditions were considered in a way which rested essentially on the comparison of different neighbouring *states*: the system was always in equilibrium. The same problem is now to be investigated by comparing a state of the system with what will be called a neighbouring *unnatural state*. Before undertaking this task it is appropriate to refer to a somewhat analogous situation which one meets, for instance, in classical mechanics when the latter is based on Hamilton's Principle. Here one has to distinguish between 'natural motions', 'unnatural motions' and 'impossible motions'. It suffices to consider a particle constrained to move on the surface of a sphere from a point A to a point B, under the action of no forces. Then the natural motion is that which takes place with constant speed along the arc of a great circle through A and B. The unnatural motion is one which takes place on the surface of the sphere but which is abstractly conceived as differing from the natural motion by virtue of the speed not being constant or the path not being the arc of a great circle; whilst, finally, an impossible motion is one which involves the necessity of the particle leaving the surface of the sphere. Thus, an unnatural motion is one which is *in fact* impossible, but the possible occurrence of which is contemplated until a certain law, viz. Hamilton's Principle, rules it out. An impossible motion, however, is one which violates the constraints, that is to say, it is considered to be impossible *a priori*.

If one argues in an appropriate fashion, a somewhat analogous situation prevails in the context of thermodynamics. Here also it is worth considering a simple example, namely just the system K_C encountered at the beginning of Section 46*a*. K_C as a whole shall be altogether isolated at all times, whilst the deformation coordinates of its subsystems K_A and K_B shall have fixed values. Suppose K_A and K_B to be initially separated from one another by an adiabatic partition. Then so long as $T'_A \neq T'_B$ the entropy of K_C is less than its entropy will be after the adiabatic partition has been removed and equilibrium has re-established itself: that this is so is just an expression of the Principle of Increase of Entropy, applied to the present situation. Now in this case the equilibrium condition $T''_A = T''_B$ is trivial in the sense that the Zeroth Law already requires it. Nevertheless it is of value to reformulate it in a manner which lends itself to suitable generalization. For this purpose one has to introduce the notion of an unnatural state. In the example under consideration this consists of the prescribed set of values of the deformation coördinates together with the temperatures T_A and T_B of K_A and K_B. Thus one pretends that K_A and K_B are separately in definite states and defines the *pseudo-entropy* S^*_C of K_C to be the sum of S_A and S_B, i.e. of the entropies of the hypothetical states of K_A and K_B. Of course T_A and T_B are not independent of each other since the isolation of K_C requires that its energy $U_C = U_A + U_B$ be fixed. This last condition is precisely analogous to the *constraint* imposed upon the motion of the particle considered above; and any hypothetical state whose energy differs from its preassigned value is an impossible state. It is convenient to take the set of unnatural states to contain one state, namely when the values of the coordinates happen to correspond to equilibrium. Thus, in the example under consideration, the unnatural state which has $T_A = T_B$ ($= T''$) is a state: its entropy is greater than that of all the unnatural states. Accordingly the state of the system may be singled out by the maximization of S^*_C.

It is not difficult to convince oneself that the maximization of the pseudo-entropy of unnatural states (when the system is isolated) is a suitable criterion for determining the actual state whenever a given system K can be modified, in the sense of Section 46*b*, to yield a system K^* such that to every unnatural state of K there correspond (true) states of the subsystems of K^*. The foregoing example of thermal conduction is typical, the modification in question being the insertion of an internal adiabatic partition. Another example is a system consisting of a mixture of chemically inert gases (see Section 82(ii)). The modified system here has the same volume and contains the same total quantity of each gas as the given system, but it incorporates an internal partition. The temperatures in the compartments on either side of it are equal and so are the pressures, but the concentrations of the constituents are not. When the partition is removed diffusion will set in, and the entropy of the final state of the original system will be greater than that of the state of the modified system.

One can construct further examples of this kind, in which the unnatural states of a given system are, as it were, 'realized' by a suitably modified system. In each case it is taken for granted that when the modification is abandoned (i.e. the original structure of the system restored) processes which will take the system to equilibrium will actually occur. Now there are situations in which it is more difficult to construct modified systems which might serve for the purpose of realizing unnatural states. A good example is that of a mixture of chemically reacting gases. In this case a modified system would be one in which somehow the constituents can be made to exist in proportions not corresponding to equilibrium. One's only hope is to suppose it to be possible to introduce an inert 'anti-catalyst' whose function is to inhibit the reaction completely. When the anti-catalyst is removed the reaction will proceed, and the Entropy Principle may be applied as before.

It will be taken for granted on the basis of the preceding discussion that it is meaningful to contemplate unnatural states in all cases of interest. Accordingly the following general principle will be adopted for the characterization of equilibria:

The entropy of any state of an isolated system is greater than the entropy of any neighbouring unnatural state. (69.1)

It should be clearly realized that the term '*unnatural state*' is not a misnomer, appearances notwithstanding; for an unnatural state is to be interpreted in terms of the state of an appropriately modified system. However, one can of course adopt (69.1) simply as a *prescription*, just as Hamilton's Principle in dynamics is a prescription for selecting natural motions. The pseudo-entropy is not a measurable property of a system, but this is equally true of the kinetic energy of the unnatural motion of a particle.

The isolation of the system is intended to imply that the deformation coordinates $\bar{x}$ have fixed values, whilst the energy has of course the same values for all unnatural comparison states. The equilibrium condition is thus, on dropping asterisks,

$$dS = 0, \quad d^2S < 0 \quad (\bar{x}, U = \text{const.}) \qquad (69.2)$$

(see also Section 70). Here S is appropriately considered as a function of $\bar{x}$, U, and of the n_i^k. In applying (69.2) one must take into account that an isolated system is closed, so that

$$\sum_{k=1}^{p} dn_i^k = 0; \tag{69.3}$$

whilst the dn_i^k must also satisfy any demands imposed upon them by the presence of chemical reactions.

It will have been noticed that (69.1) refers to *neighbouring* unnatural states of a given state. It may happen that there are several solutions satisfying (69.2) so that there are several possible equilibria; one of these is 'absolute', the other are 'relative'. In practice this means that when a system is disturbed from one of these equilibria it may end up in one of the others, depending on the extent to which it was disturbed. In such a situation the Entropy Principle must of course again be satisfied.

(*b*) The condition $dS = 0$ yields explicit conditions of just the form encountered in Section 68, so that one is not confronted with any inconsistencies arising from the adoption of alternative points of view. Recalling (67.9) one has

$$TdS = dU + PdV - \Sigma\Sigma\mu_i^k dn_i^k, \tag{69.4}$$

so that for equilibrium one must have, since $dU = dV = 0$

$$\Sigma\Sigma\mu_i^k dn_i^k = 0. \tag{69.5}$$

This is formally identical with (68.2). Here, however, the dn_i^k are the differences between the values of the internal coordinates corresponding to the natural (internal) state on the one hand and some unnatural state on the other. The dn_i^k are thus not brought about by changes of external coordinates, but are in effect independent of each other, except to the extent that they must satisfy those equations of constraint which will assure any comparison state to be unnatural, but not impossible. By way of example, if one has a single, closed, chemically active phase, the dn_i need only satisfy

$$dn_i = \sum_r \nu_i^{(r)} d\xi^{(r)}, \tag{69.6}$$

(cf. (68.13)), where the $d\xi^{(r)}$ may be chosen arbitrarily straight away; and then (68.16) follows at once.

In place of $S(V, U, n_1^1, \ldots, n_z^p)$ other characteristic functions may be used equally well. Just as (68.2) arose from the inequality $\Delta S \geqslant 0$ for an isolated system, so (50.12), i.e.

$$\Delta S \geqslant \int dQ/T, \tag{69.7}$$

is available under other conditions. For example, suppose that all unnatural states of the given closed system K are taken at fixed pressure and temperature. This means that the modified system K^* is not adiabatically isolated as a whole but that it is now in interaction with its surroundings in such a way that the latter are at a fixed temperature T throughout and exert a fixed pressure P on K^* at all times. Then $\Delta S \geqslant Q/T$, whence, comparing the Gibbs function of K^* with that of the final state of K after its internal structure has been restored

$$\Delta G = \Delta U + P\Delta V - T\Delta S = Q - T\Delta S \leqslant 0. \tag{69.8}$$

Thus the maximization of S is now replaced by the minimization of G:

$$dG = 0, \quad d^2G > 0 \quad (P, T = \text{const.}). \tag{69.9}$$

This of course merely gives (69.5) again, as it must, since for equilibrium as such the form in which fixed external conditions are specified is obviously irrelevant. Nor is it of any account that in principle the μ_i^k are regarded as functions of V, U in the context of (69.2), but of P, T in (69.9). The freedom one has in the choice of a particular characteristic function evidently can have no substantial consequences unless one is concerned with questions in which not merely neighbouring states are involved, i.e. when one interests oneself in the behaviour of a system with regard to several relative equilibria which may be possible. Physically, this means that a large disturbance of, say, an adiabatically isolated system on the one hand, and, on the other, of a system capable of interacting thermally with its surroundings, may have different consequences.

70. Stability

In the preceding section the equilibrium conditions such as (69.2) or (69.9) consisted of two parts, the first of which was in each case the condition for the existence of an extremum of some kind, whilst the second then fixed the character of the extremum. Thus one might well be able to find values of the coordinates $\bar{x}, U, n_1^1, \ldots, n_z^p$ such that $dS = 0$, but such that $d^2S > 0$ for some set $d\bar{x}, dU, dn_1^1, \ldots, dn_z^p$. There then exists a sequence of unnatural states of increasing entropy in a neighbourhood of the state in question, and, in accordance with previous assumptions, the latter is therefore not a possible state of the system. Consequently, having solved the equations arising from the extremum condition one must always verify that one has arrived at an extremum of the required character. It is usual to refer to the conditions $d^2S < 0$, $d^2G > 0, \ldots$ as *stability conditions*.

The demonstration that uniformity of temperature within a homogeneous substance satisfies the condition of stability is instructive, even if it is redundant, The unnatural states are realized by a modified system K^* containing an internal rigid adiabatic partition. The two parts of K^* are distinguished by subscripts A and B as usual. Their specific heats (at constant volume) shall be c_A and c_B. Let derivatives with respect to U (at constant volume) here be distinguished by primes, e.g. $\partial S_A/\partial U_A = S'_A$. Then, for equilibrium,

$$dS = S'_A dU_A + S'_B dU_B = 0. \tag{70.1}$$

However, $dU_A = -dU_B$ and $S' = T^{-1}$, whence (70.1) gives (of course!) $T_A = T_B$ (= T, say). Next

$$d^2S = \tfrac{1}{2}[S''_A(dU_A)^2 + S''_B(dU_B)^2] = \tfrac{1}{2}(S''_A + S''_B)(dU_A)^2.$$

Since $S''_A = (1/T_A)' = -(c_A T_A^2)^{-1}$, etc., one has finally

$$d^2S = -(1/2T^2)(c_A^{-1} + c_B^{-1})(dU_A)^2 < 0, \tag{70.2}$$

since the specific heats are positive.

Instead of modifying the system above through the introduction of a rigid adiabatic partition, one may alternatively take the partition to be diathermic but movable, the entire system having a fixed volume and temperature, the latter being achieved by thermal interaction with the surroundings. The appropriate characteristic function is F, which is to be minimized. Then

$$dF + d^2F = \sum_{A,B} (F_{,V} dV + \tfrac{1}{2} F_{,VV} dV^2). \tag{70.3}$$

Recalling that $F_{,V} = -P$, $dV_A = -dV_B$, the equation corresponding to (70.1) gives $P_A = P_B$, whilst that corresponding to (70.2) yields the condition

$$P_{,V} < 0. \tag{70.4}$$

That (70.4) must hold for K_A and K_B separately (and so for K as a whole) rests on the possibility of taking the internal partition to divide K into parts the ratio of the volumes of which may be prescribed arbitrarily. The same remark would allow one to conclude that c_A and c_B are separately positive if one did not know this already. It should be noted that in deriving (70.4) K was *assumed* to be homogeneous. However, one must bear the possibility of phase changes in mind, so that neighbouring unnatural states in which the given substance is distributed over different phases should be contemplated (cf. Section 85*b*).

As regards physico-chemical equilibrium, stability requires that

$$\sum_{k,l}\sum_{i,j} (\partial^2 G/\partial n_i^k \partial n_j^l)\, dn_i^k\, dn_j^l > 0 \tag{70.5}$$

for all unnatural states near the state satisfying (69.5), P and T having fixed values. Thus, the quadratic form on the left of (70.5) must be positive definite, the dn_i^k being subject to the usual constraints which ensure that the comparison states are not impossible. With the help of the equations of constraint one may bring (70.5) into the generic form

$$\Sigma \Gamma_{ij}^{kl}\, dn_i^k\, dn_j^l > 0, \tag{70.6}$$

where the summation now only goes over those values of the indices for which the corresponding n_i^k may be varied independently. The usual algebraic conditions ensuring the positive definite nature of a quadratic form may then be applied to (70.6); see Section 89.

There may of course be special situations requiring separate investigation. For example, it may happen that S has an extremum such that d^2S and d^3S vanish identically, whilst d^4S does not. Then this extremum can still be a maximum provided the *quartic* differential form d^4S is negative definite. However, such special cases can be dealt with as they arise, and the remarks of this section will suffice as far as general principles are concerned.

CHAPTER 9

MISCELLANEOUS TOPICS

71. Ideal and super-ideal gases

(*a*) The inclusion of a discussion of ideal gases in this chapter rather than in the next is somewhat arbitrary, and it is motivated, at least in part, by the prominent position they occupy in applications of the general theory. Over and above this, there are certain issues of principle at stake, and it is proper to elucidate these at an early stage. In this respect the main question concerns the definition of an ideal gas. Accordingly, an ideal gas, unqualified, will be understood to be a substance, i.e. a fluid, which obeys Boyle's Law *and* whose energy depends upon the temperature alone:

$$PV = \phi(t), \quad U_{,V} = 0, \tag{71.1}$$

where t is an empirical temperature. Whether such fluids are actually to be found in nature is irrelevant. However, as observed in Section 48*a*, provided V and PV are sufficiently large, real gases obey (71.1) quite closely. Drawing in this section upon some identities yet to be derived, (81.3) shows immediately that

$$P = T\theta(V),$$

and comparison with the first of (71.1) then shows that the product PV is proportional to the absolute temperature. With the understanding that one is dealing with one mole of the gas in question, the constant of proportionality is usually written as R, so that (71.1) is equivalent to

$$PV = RT, \tag{71.2}$$

as was already shown in Section 48*a*. U is some function of T only, say

$$U = f(T), \tag{71.3}$$

which need not be linear. Indeed under appropriate conditions the behaviour of many real gases approximates (71.2) quite closely, without $f(T)$ being even remotely linear in the range 100–1000 °K say.

From (71.2) and (71.3) one easily finds that

$$S = \int T^{-1} f_{,T}(T)\,dT + R\ln V + \text{const.} \tag{71.4}$$

Now (71.4) is clearly in conflict with the Third Law. It is true that provided $f_{,T}$ goes to zero in a suitable way S remains finite as $T \to 0$; but evidently $S_{,V}$ does not tend to zero as $T \to 0$, and (60.2) is violated. Accordingly, at sufficiently low temperatures no real gas can even approximately behave in conformity with (71.1). The question arises as to how the latter might be suitably modified. This can evidently hardly be done on a purely phenomenological level. Still, it is worth seeing what can be achieved.

At ordinary temperatures and pressures the specific heats at constant volume of the inert gases are observed to be $3R/2$, so that for these $PV = 2U/3$. (The additive constant in the relation $U = 3RT/2 + \text{const.}$ has been dropped since it does not affect subsequent results.) With regard to the common diatomic gases such as H_2, O_2, CO, ... the specific heat is, under like conditions, nearly $5R/2$. That of H_2 falls off to $3R/2$ as T decreases to about 40 °K, but at this temperature the other gases have already liquefied, that is to say, the effects of their imperfections makes it useless to consider them in the present context. However, there is a suggestion here that it might be worth investigating the consequences of defining a class of substances, called *super-ideal gases*, by the requirements that:

(i) they obey the relation

$$PV = 2U/3, \tag{71.5}$$

(ii) for sufficiently high temperatures they behave as ideal gases,

(iii) they do not violate the Third Law.

Whether these conditions are mutually compatible of course remains to be seen.

The motivation of (71.5) as set out above is hardly convincing. However, one may argue as follows. Equation (71.5) certainly holds very nearly for some gases at ordinary temperatures, gases which are (approximately) ideal. Now the definition of an ideal gas, in effect (71.2), together with the constant value of $3R/2$ of the specific heat c gives rise to (71.5), but conversely (71.5) does not necessarily imply (71.2) or the constancy of c; so that it is somewhat weaker. An important aspect of (71.5) is now that it makes no explicit reference to the temperature; for which reason one may hope to be able to accommodate condition (iii). (That the asymptotic condition (ii) can be satisfied is obvious from the fact that ideal gases with $c = 3R/2$ satisfy (71.5).) As a matter of fact, the absence of the temperature from (71.5) to some extent lends this relation a purely mechanical character. By way of illustration, one may consider the isotropic motion of N structureless particles in a rigid box of volume V, assuming there to be no mutual interactions between them. It is known from entirely elementary considerations that

$$\left.\begin{aligned} E &= Nm\langle c^2[(1-v^2/c^2)^{-\frac{1}{2}}-1],\rangle \\ 3PV &= Nm\langle v^2(1-v^2/c^2)^{-\frac{1}{2}}\rangle. \end{aligned}\right\} \tag{71.6}$$

Here E is the total kinetic energy of the system, m and v the rest mass and speed of a particle, whilst c is the speed of light. The broken brackets denote averages over the speeds. When all speeds are much less than c one infers that

$$PV = 2E/3, \tag{71.7}$$

whereas when all speeds are equal to, or very nearly equal to c,

$$PV = E/3. \tag{71.8}$$

No thermodynamic concepts are involved in these results, whether phenomenological or otherwise; and if one insists on not considering particles one may instead contemplate the mean rate of transport of momentum across a surface by 'beams' of some fluid.

After these heuristic remarks, let (71.5) be taken for granted. To study its consequences it is convenient to take in its place the slightly more general relation

$$PV = rU, \tag{71.9}$$

where r is a positive constant. From (59.3), with $x = V$, one then has the differential equation

$$VP_{,V} - rTP_{,T} + (r+1)P = 0. \tag{71.10}$$

One solution must be $P = RT/V$. Accordingly set

$$P = RV^{-1}T\phi(V, T),$$

so that (71.10) becomes

$$V\phi_{,V} - rT\phi_{,T} = 0. \tag{71.11}$$

The solution of this is, virtually by inspection, $\phi = g(TV^r)$, where g is an arbitrary function of its argument. Thus, finally,

$$PV = RTg(z), \quad z = TV^r. \tag{71.12}$$

This, then, is the generic equation of state. The entropy is obtained in the usual way, and is given by

$$S = \bar{R}\int (g' + z^{-1}g)\,dz, \tag{71.13}$$

where $\bar{R} = R/r$, and a prime denotes differentiation with respect to z. Also

$$c = \bar{R}(zg)'. \tag{71.14}$$

For the substance to be a super-ideal gas $g(z)$ must satisfy certain conditions. Thus (i) it must tend to unity as z becomes sufficiently large, and (ii) as $z \to 0$ it must behave in such a way that S tends to a finite limit. With regard to the second condition there is no *a priori* reason for assuming that near $z = 0$ g can be expanded in *integral* powers of z. Let it therefore be supposed merely that $g(z)$ can be expanded in ascending powers of z when z is sufficiently small. To a term of the form az^α in the expansion there corresponds in S a term $[a(\alpha+1)/\alpha]z^\alpha$ if $\alpha \neq 0$, or $a\ln z$ if $\alpha = 0$. Hence the permissible values of α are $\alpha = -1$ and $\alpha > 0$:

$$g(z) = az^{-1} + bz^\beta + O(z^\gamma) \quad (\gamma > \beta > 0). \tag{71.15}$$

Then

$$c/\bar{R} = b(\beta+1)z^\beta + O(z^\gamma). \tag{71.16}$$

In the statistical theory the so-called ideal Fermi–Dirac and Bose–Einstein gases are indeed super-ideal, and they obey (71.12) and (71.15, 16), with $r = \frac{3}{2}$. Moreover, for the Fermi–Dirac gas $a \neq 0$, $\beta = 1$, whilst for the Bose–Einstein gas $a = 0$, $\beta = \frac{3}{2}$.

(*b*) Under normal conditions a gas, regarded as a closed system, has a mass which is independent of the coordinates. One may, however, envisage circumstances under which its mass increases substantially, namely when relativistic effects make themselves felt. Such conditions were here not taken into account and the super-ideal gas described above should therefore strictly speaking be qualified as being non-relativistic. Actually the qualification is academic in practice, except in astrophysical situations. On the other hand, consider electromagnetic radiation within an otherwise empty material enclosure at temperature T. This radiation is called black-body radiation when it is in equilibrium with the enclosure. It is not difficult to demonstrate by thermodynamic reasoning that the character of the radiation is determined by T alone. In particular the radiation is isotropic and its energy-density $u\,(= U/V)$ depends upon T alone. Given only the isotropy, electrodynamic theory leads to the result that

$$u - 3P = 0, \tag{71.17}$$

i.e. (71.9) is obeyed, with $r = \frac{1}{3}$. [It is instructive to recall (71.8) here.] From (71.12), then, u will depend upon V unless $g(x)$ is proportional to x^3. Thus

$$u = aT^4, \tag{71.18}$$

where a is a constant; which is, in effect, *Stefan's Law.*

It will be realized that the results of this section are not 'purely' thermodynamic, to the extent that they rest upon a specific form of the caloric equation of state: a purely thermodynamic result would be a conclusion which follows merely from the knowledge that such an equation of state exists at all.

72. Behaviour under Lorentz transformations

Hitherto, whatever the system K under consideration may have been, it has always been supposed to be at rest in the frame of reference A_0 (proper frame) in which the various measurements are carried out. Granted that A_0 is inertial, one may now introduce a second inertial frame A which moves relatively to A_0 with uniform speed u. It suffices to take the two frames to be so arranged that space-time measurements in them are related to one another by the standard Lorentz Transformation

$$x = \gamma(x_0 + ut_0), \quad y = y_0, \quad z = z_0, \quad t = \gamma(ux_0 + t_0), \tag{72.1}$$

where $\gamma = (1 - u^2)^{-\frac{1}{2}}$, and units have been so chosen that the value of the speed of light is unity.

Attention will be confined to the case where K is a homogeneous fluid in an enclosure of proper volume V_0. The problem is to determine how the measures of the usual thermodynamic quantities referred to A_0 are related to their measures when referred to A. For this purpose, it will be accepted that (i) *the speed u of K relatively to A_0 may be regarded as an additional thermodynamic* *coordinate*; and (ii) *the basic laws of thermodynamics continue to hold unchanged.* Of course, when u is varying in time, (72.1) has to be understood as the connection between A_0 and the instantaneous proper frame of K.

A knowledge of purely mechanical theories will, as hitherto, be presupposed. It is a standard result of the Lorentz covariant form of the dynamics of continuous systems that if $\mathbf{u}$ is the three-velocity of K then $(U+PV)\mathbf{u}$ $(=H\mathbf{u})$ is its momentum and the four quantities $p^k = (H\mathbf{u}, H)$ form the components of a four-vector. K therefore has the same momentum as a particle whose rest energy is H_0. (The additional term $P_0 V_0$ over and above U_0 allows for the momentum density which has to be associated with moving stresses.)

Only the x-component of momentum of K is non-zero, and it is

$$M = \gamma u H_0. \tag{72.2}$$

From the transformation properties of p^k it follows that

$$U = \gamma(U_0 + u^2 P_0 V_0) = \gamma H_0 - \gamma^{-1} P_0 V_0, \tag{72.3}$$

taking into account that the pressure is a scalar,

$$P = P_0, \tag{72.4}$$

and that, because of the Lorentz contraction,

$$V = \gamma^{-1} V_0. \tag{72.5}$$

Contemplate now a quasi-static transition in which the thermodynamic coordinates, including the speed u of K, change at an infinitesimal rate. Even in a change of u alone, if the latter occurred at a finite rate the transition would not be reversible. The finite acceleration of K would produce relative displacements of the elements of the fluid, leading to irreversibility just as much as they lead to irreversibility when they occur as a consequence of non-static changes of volume when K is at rest in its terminal states. If the transition is between neighbouring states the work $-dW$ done on K will be given by

$$dW = PdV - udM, \tag{72.6}$$

the second term representing the work done by the force causing the change of speed of K. If the transition is adiabatic $dU = -dW$, in view of the definition of the energy of K. Then, as usual,

$$dQ = dU + PdV - udM. \tag{72.7}$$

Using (72.2–5) this becomes

$$\begin{aligned} dQ &= d(\gamma H_0) - \gamma^{-1} V_0 dP_0 - u\, d(u\gamma H_0) \\ &= \gamma^{-1}(dU_0 + P_0 dV_0) + H_0[d\gamma - u\, d(u\gamma)]. \end{aligned}$$

The last term, however, vanishes identically and one is left with

$$dQ = \gamma^{-1} dQ_0. \tag{72.8}$$

Now S will in principle be a function of all the coordinates. However, by inspection of (72.8), the differential du is absent from dS. It follows that S is independent of u, i.e. *S is invariant under Lorentz transformations*; and at the same time one must have

$$T = \gamma^{-1} T_0. \tag{72.9}$$

73. Systems in the terrestrial gravitational field

(*a*) All laboratory experiments of necessity have to be carried out in the presence of the earth's gravitational field. The effects of this field are now to be considered very briefly. One recognizes at once that the *restriction to the laboratory scale* implies a situation vastly simpler than that which would obtain if systems on an astronomical or cosmic scale were to be contemplated. Thus it is entirely sufficient: (i) to take the gravitational field to be Newtonian, i.e. to be described by a single scalar potential ϕ, and not by the ten potentials g_{ik} of the general theory of relativity; (ii) to regard the field as external and fixed, i.e. independent of the structure and state of any thermodynamic system K situated within it; and (iii) to take the level surfaces of ϕ to be a set of mutually parallel planes. In terms of a suitably orientated Cartesian set of coordinates x, y, z the potential is then a linear function of z alone:

$$\phi = gz + \text{const.}, \tag{73.1}$$

where g is a constant.

It is evident that now K can never be homogeneous, for the pressure P at any rate will be a function of z. This being so, one has a verbal difficulty, for a phase was previously defined to be a physically homogeneous part of K which possessed a sharp boundary; so that (very special cases apart) one could have at most one gaseous phase. Suppose, indeed, that K is just a single gas contained in an enclosure of volume V. Then one certainly has no sharp boundary anywhere, yet the gas is not homogeneous. Moreover, if one has a mixture of gases the composition of the mixture will be a function of z. One therefore has apparently no option but to think of K as stratified into layers of thickness δz, the boundaries of the layer at z_1 coinciding with the planes $z = z_1$ and $z = z_1 + \delta z$, and regarding each such layer as a separate phase. The number of phases is thus infinite. With this terminology one would in effect make nonsense of the phase rule (Sections 87–88) for instance. Clearly one would still wish to be able to speak of a gas, for example, as constituting a single 'phase'. Accordingly, it will be supposed that when K is situated in a gravitational field one can still distinguish between various *phases*, each phase being a part of K with a sharp physical boundary (so that one can again have only one gaseous phase), whilst two phases are distinct in the sense in which two different modifications of ice are distinct. Any particular phase then divides itself into (an infinite number of) *metaphases*, differing amongst each other with regard to the local state in a certain way.

Consider, then, the case where K consists of n moles of a single constituent C of molar mass M within an enclosure of volume V, only one phase being present. Each metaphase is characterized by a particular value of z, and one may refer to it as 'the metaphase at z'. One has to realize from the start that in an infinitesimal transition of K the work done by it depends upon the particular way in which the enclosure is being deformed, since the pressure exerted on different elements of the enclosure depends upon the value of z which specifies the location of the element. Now imagine a small amount dn of C to be moved quasi-statically from the metaphase at z_1 to the metaphase at z_2. This might be done by having a narrow tube which leads into K at z_1 and z_2, there being a thin freely movable membrane within it. The pressures on the two sides of the membrane are equal, and when it is moved at an infinitesimal rate no work has to be done on it. Now the work done *by* K when the amount dn leaves the metaphase at z_1 is

$\mu(z_1)\,dn$, where μ is the *usual* chemical potential of C. (The notation notwithstanding, this depends on z_1 only because the thermodynamic coordinates do so, i.e. it does not depend on z_1 *explicitly*.) Similarly, the addition of dn at z_2 requires an amount of work $-\mu(z_2)\,dn$ to be done by K. On the other hand, the whole system has to all intents and purposes not changed at all, so that $(\mu(z_1)-\mu(z_2))\,dn$ must be just equal to the amount of work which has to be done to move the mass $M\,dn$ against the gravitational field from z_1 to z_2, i.e. $M(\phi(z_2)-\phi(z_1))\,dn$. Recalling that z_1 and z_2 may be chosen arbitrarily, one therefore has throughout K

$$\mu^* \equiv \mu(P, T)+M\phi(z) = \text{const.} \tag{73.2}$$

The appearance of μ^* simply amounts to this: that if any matter added to K is taken to have come from some standard position in the gravitational field, then the addition of an amount dn requires work

$$dW = (\mu+M\phi)\,dn = \mu^*\,dn, \qquad (73{\cdot}3$$

to be done, for the mass $\mu\,dn$ must first be brought from the standard position to the required position. In effect, therefore, the presence of the field is taken into account by replacing μ by μ^* throughout, and regarding every phase as the sum total of its metaphases. For a multiphase system of several constituents $C_1, C_2, \ldots$, of molar masses $M_1, M_2, \ldots$ one thus has

$$G = \sum_k \sum_i \int \mu_i^{k*}\,dn_i^k, \tag{73.4}$$

where dn_i^k is the mole number of C_i in a typical metaphase of the kth phase. (This result will obviously continue to hold in an inhomogeneous field.)

The temperature is constant throughout K, as usual. This is perhaps not quite as obvious as hitherto, since different parts of K are now distinguished by their different positions in the field. (Indeed, one would expect this result to hold strictly only in the non-relativistic limit, since flow of energy is accompanied by, or better, equivalent to a flow of inertial, and therefore of gravitational mass. In a fully relativistic theory the presence of a gravitational field will therefore fundamentally affect previous results.) However, one may simply consider unnatural states for which the temperature is not uniform and so infer the stated result from the Entropy Principle, as in Section 70. Further, (73.2) holds separately for each constituent of a phase:

$$\mu_i^* = \mu_i + M_i\phi = \text{const.} \tag{73.5}$$

μ_i is a function of P, T and the concentrations, whilst P and ϕ are functions of z.

In the case of a single constituent, differentiating (73.5) with respect to z,

$$VP_{,z} + M\phi_{,z} = 0,$$

where V is the molar volume. However, taking (73.1) into account, and since $M/V = \rho$ is the (material) density, this becomes

$$P_{,z} = -g\rho, \tag{73.6}$$

the usual equation of hydrostatic equilibrium. In the case of an ideal gas this gives

$$P = \text{const.} \exp(-gMz/RT). \tag{73.7}$$

If one uses (82.16), relating to a mixture of ideal gases, one can show in much the same way that

$$c_i = \text{const.} \exp(-gM_iz/RT). \tag{73.8}$$

(*b*) Even when one has simply a gas within an enclosure of volume V there is no longer any general way of prescribing a choice of deformation coordinates, as already remarked. When the field is absent one will usually choose V. In the presence of a field, however, one is confronted with an infinity of potential deformation coordinates, if one considers K as a whole. As a simple example, let the enclosure be a right circular cylinder of cross-sectional area A and height h, whose (plane) ends coincide with level surfaces of ϕ. The volume V may be changed, for instance, by depressing the upper surface of the cylinder, or by deforming the curved side; and the corresponding amounts of work done will in general differ from each other. For each possibility of changing V one therefore has to contemplate a separate deformation coordinate. How one does this in detail has to be decided from case to case.

One is accustomed to thinking of the internal energy of a system containing a perfect gas as not depending upon the volume. This, indeed, seems to be a matter of definition. The purpose of this final remark is to show that one has to be careful in this regard when a gravitational field is present. Contemplating the example above, let it be supposed that the cylinder contains just one mole of a perfect gas of constant specific heat $3R/2$, for the sake of argument. This constancy refers of course to a gravitation-free situation. Further let it be supposed that the volume of the cylinder can be changed only by moving its upper surface, so that one may take just h as the only deformation coordinate in this case. Then one has

$$U = A\int_0^h (\tfrac{3}{2}RT/M+gz)\rho\, dz, \quad M = A\int_0^h \rho\, dz. \qquad (73.9)$$

Recalling that $P = \rho RT/M$, $T = \text{const.}$, these may be evaluated by using (73.7). After some manipulation one gets the result

$$U = RT\left(\frac{5}{2}-\frac{\xi}{e^{\xi}-1}\right), \quad \xi = Mgh/RT. \qquad (73.10)$$

Obviously U depends upon h, whilst, incidentally, the specific heat varies from $5R/2$ to $3R/2$ as T goes from values much less than Mgh/R to values much greater than Mgh/R. The entropy is here

$$S = R\{\tfrac{5}{2}\ln T+\ln(1-e^{-\xi})-\xi(e^{\xi}-1)^{-1}+\text{const.}\}. \qquad (73.11)$$

For ordinary gases under normal conditions $\xi \ll 1$, so that the gravitational field may be disregarded. In meteorological applications, for instance, significant values of ξ may, however, occur. (Reference may also be made to Section 85*c*.)

74. Remark on surface effects

Hitherto all surface effects were regarded as negligible. This point of view is justified whenever any system or phase is sufficiently bulky, for the ratio of surface area to volume of a body tends to zero as its linear dimensions tend to infinity. For phases of finite extent, however, one must on occasion take surface effects explicitly into account. This section accordingly deals with a few relevant points of principle, detailed considerations being left aside, as these would lead too far afield.

The boundary between two phases (like the boundary of a phase where it is in contact with an enclosure) was, by definition, a sharply defined surface. Further, the phase was understood to be homogeneous throughout its interior, gravitational effects being now disregarded. This picture, however, does not conform strictly to the physical situation. Two phases in contact are in fact separated by an inhomogeneous transition layer, often called a 'surface phase'. On the face of it this terminology leaves something to be desired. In the first place, there is some difficulty in deciding where the 'boundaries' of the transition layer are situated; whilst, since these boundaries are, in the nature of things, not sharply defined, the transition layer is not a phase in the ordinary sense of the term in any case. At best one could think of it as a continuous sequence of metaphases.

The simplest procedure is then the following. One *defines* the boundary between two phases to be a certain surface Σ, the location of which is determined by a prescription. (In every case it will of course coincide very closely with the macroscopically visible boundary.) Thus, from amongst the $z+1$ components $C_0, C_1, \ldots, C_z$ present in the phases choose a particular one, say C_0. The components are supposed to be numbered in such a way that the volume concentrations $'n_0^k$ of C_0 at points within the bulk of the two phases are not equal. The combined volume V of the phases and the total amount $\bar{n}_0$ of C_0 present in them are given. Then Σ is to be taken in such a way that

$$V^1 + V^2 = V, \quad 'n_0^1 V^1 + 'n_0^2 V^2 = \bar{n}_0,$$

where V^1 and V^2 are the volumes of the phases corresponding to the position of Σ thus defined. As far as C_0 is concerned, the location of Σ is just that which would obtain if the volume concentrations of C_0 were uniform throughout the phases.

Let $'n_i^k$ $(i \neq 0)$ be the bulk volume concentrations of the other components. If the phases were homogeneous the total amount of C_i present would be $\bar{n}_i = 'n_i^1 V^1 + 'n_i^2 V^2$. In fact it will be greater than this by an amount $\Delta\bar{n}_i$, which may of course be negative. Then $\Delta\bar{n}_i$ may be formally associated with the surface, and it suffices here to take it as uniformly distributed over Σ. Next, let U be the joint energy of the two phases, which is in principle obtained by direct measurement as usual. Further let U^k $(k=1,2)$

be the energies of the separate phases, calculated as if they were entirely homogeneous. Then one can associate with Σ an energy

$$U^{\sigma} = U - U^1 - U^2; \tag{74.1}$$

and an entropy S^{σ}, Helmholtz function $F^{\sigma}, \ldots$, associated with Σ, may be defined after the same fashion.

In this way Σ becomes a thermodynamic system in its own right. Its extent is its area A^{σ}, whilst its composition is described by the numbers $n_i^{\sigma} = \Delta \bar{n}_i$. It is *this* system which may be called the *surface phase* associated with the given phases. In a multiphase system one may have several surface phases, in which case one may simply regard σ as an index which numbers the various surface phases.

It should be noted that the presence of surface effects in a multiphase system does not imply a breakdown of the results of Section 24 concerning the additivity of energy. In the earlier situation each subsystem retained its identity in the compound system, that is, when two subsystems at the same temperature were brought into mutual contact, neither of them was affected. On the other hand, suppose each subsystem to contain a phase in a state such that in the hypothetical absence of surface effects the phases would be in equilibrium when brought into direct mutual contact. Then the removal of the partition between them represents an essential *modification* of the system K as a whole, not so much because the states of the subsystems will be affected, but rather because a further deformation coordinate must be associated with K, which may be taken to be the area A of Σ. The additivity of energy, i.e.

$$U = \sum_k U^k + \sum_{\sigma} U^{\sigma},$$

survives now *by definition*, that is to say, as a consequence of the definition (74.1).

Generally, if A^{σ} is the area of the surface phase Σ^{σ},

$$T^{\sigma} dS^{\sigma} = dU^{\sigma} - \gamma^{\sigma} dA^{\sigma} - \sum_i \mu_i^{\sigma} dn_i^{\sigma}. \tag{74.2}$$

Here μ_i^{σ} is the chemical potential of C_i in the surface phase Σ^{σ}, whilst γ^{σ} is the *surface tension* of Σ^{σ}, which is the two-dimensional analogue of the three-dimensional pressure $-P$. In this kind of relation a surface phase is indeed treated as if it were physically quite a separate system. This is conceptually convenient even if it does not quite correspond to the actual state of affairs.

With regard to the equilibrium of a multiphase system K, it is hardly necessary to go into a detailed discussion of equilibrium conditions. Once again one applies the Entropy Principle to a set of unnatural states. One concludes in the first place that the temperature is uniform throughout K, as expected. Further the chemical potentials of any one constituent is the same in all phases $k, \ldots, \sigma, \ldots$ in which it is present:

$$\mu_i^k = \mu_i^l = \mu_i^\sigma = \mu_i^\tau. \tag{74·3}$$

Again, if Σ^σ separates phases k and l, then $P^k = P^l$ provided Σ^σ is plane. When it is not, $P^k - P^l$ depends on the principal radii of curvature of Σ^σ. In particular, when the phase l is a spherical bubble of radius r, immersed in the phase k, one has

$$P^k - P^l = -2\gamma^\sigma/r. \tag{74·4}$$

It may be noted in passing that stability requires $\gamma^\sigma \geqslant 0$.

Characteristic functions for K will now depend on further variables. One has an even greater freedom in the choice of independent variables than before, so that a multitude of such functions presents itself. One such function is $G(P, T, A^\sigma, n_i^k, n_j^\tau)$, where the generic indices go over all their possible values. Of whatever independent variables G is regarded as a function, one will have

$$G = \sum_k \sum_i \mu_i^k n_i^k + \sum_\sigma \sum_i \mu_i^\sigma n_i^\sigma. \tag{74·5}$$

On the whole, then, though the inclusion of surface effects brings with it greater complexity in detail, the theory is not substantially altered with regard to its general formal aspects.

75. Thermal conduction. Localization of energy and of entropy production

(*a*) Certain aspects of the problem of thermal conduction were already discussed in Section 19*c*. It is instructive to enlarge a little upon these remarks, with particular reference to the Entropy Principle. Recall the conceptual device of instantaneously subdividing by means of adiabatic partitions the given system K^* into a sufficiently large number of subsystems, or 'cells', each of volume $d\mathfrak{v}$, say. For the sake of simplicity, the substance of which K^* consists will be supposed to be absolutely rigid, so that its density is independent of the temperature, whilst K^* then also cannot do any work on its surroundings. Any state of the subdivided system $K^\dagger$ is then a set of values of the temperatures within all its cells.

The energy of $K^\dagger$ is the sum of the energies of all its cells:

$$U = \Sigma u\, d\mathfrak{v}. \tag{75.1}$$

Here u is the energy density (i.e. energy per unit volume) of a cell. The entropy is

$$S = \Sigma s\, d\mathfrak{v}, \tag{75.2}$$

where s is the entropy density. Imagine all the partitions to be removed for a small time δt only. When equilibrium has set in again the final values of the energy and entropy of $K^\dagger$ differ from their initial values by amounts

$$\delta U = \Sigma \delta u\, d\mathfrak{v}, \quad \delta S = \Sigma T^{-1} \delta u\, d\mathfrak{v}, \tag{75.3}$$

since the change in the temperature in a typical cell is δT, and the relation $\delta S = \delta u / T$ is one between *functions*, so that the irreversibility of the process, considered from the point of view of $K^\dagger$ as a whole, is irrelevant. Next, one may imagine the partitions to be removed repeatedly over small times δt at intervals widely separated in time. On each occasion the various equations above will apply.

The sequence of states of $K^\dagger$ is evidently a sequence of unnatural states of K^*. *The formalism of thermodynamics may now be extended* beyond its classical boundaries by postulating that, when thermal conduction takes place, K^* can be regarded as passing through a continuous sequence of unnatural states, so long as in any cell the temperature can be defined as in Section 19*c*. [Operationally this means that the reading of a 'local' thermometer must not depend on its construction, and if one imagines at any stage the cell in which the thermometer is situated to be adiabatically isolated, the reading must not sensibly change as a consequence.] Thus, it is now accepted that in the equations (75.3) $\delta u / \delta t$ and $\delta s / \delta t$ may be replaced by time derivatives $\dot{u}$ and $\dot{s}$. At the same time it shall be permitted to replace the summations by integrations over the volume $\mathfrak{V}$ of K^*:

$$\dot{U} = \int \dot{u}\, d\mathfrak{v}, \quad \dot{S} = \int T^{-1}\dot{u}\, d\mathfrak{v}. \tag{75·4}$$

T is supposed to be a continuous, differentiable function of fixed Cartesian coordinates x_1, x_2, x_3, $(=x_i)$. [These x_i must of course not be confused with external thermodynamic coordinates.] One can always determine a vector $\mathbf{q}$ such that

$$\dot{u} = -\operatorname{div}\mathbf{q}. \tag{75·5}$$

Then

$$\dot{U} = -\oint \mathbf{q}\,.\,d\mathfrak{f}, \tag{75·6}$$

where the integral extends over the surface $\mathfrak{F}$ of K^*. Now suppose K^* as a whole to be adiabatically isolated. Then $\dot{U} = 0$, and for this to be the case it is sufficient that $\mathbf{q}$ should vanish over $\mathfrak{F}$. Further

$$\begin{aligned}\dot{S} &= -\int T^{-1}\operatorname{div}\mathbf{q}\, d\mathfrak{v} \\ &= -\int [\operatorname{div}(T^{-1}\mathbf{q}) + T^{-2}\mathbf{q}\,.\operatorname{grad} T]\, d\mathfrak{v}.\end{aligned}$$

The first term of the integral gives rise to a surface integral which vanishes, since $\mathbf{q} = 0$ over $\mathfrak{F}$. One is left with

$$\dot{S} = -\int T^{-2}\mathbf{q}\,.\operatorname{grad} T\, d\mathfrak{v}. \tag{75·7}$$

So far $\mathbf{q}$ is determined only by (75·5) in the interior of $\mathfrak{V}$. Now when T is constant in a region however small, $\dot{u} = 0$, and therefore one may assume that $\mathbf{q} = 0$ when $\operatorname{grad} T = 0$. (This assumption will, however, be examined more closely below.) For a start, suppose K^* to be isotropic. Then at any point the only distinguished direction is that of $\operatorname{grad} T$. Further, $\mathbf{q}$ and $\operatorname{grad} T$ are both vectors and so one must have

$$\mathbf{q} = -k \operatorname{grad} T, \tag{75·8}$$

where k is a scalar. k need not be constant but may depend on x_1, x_2, x_3 and on T and its derivatives. As it is already known that energy is transferred from higher to lower temperatures (Section 26) it follows that

$$k \geqslant 0. \tag{75·9}$$

Hence (except in the trivial case in which $k = 0$ everywhere)

$$\dot{S} = \int k T^{-2} |\operatorname{grad} T|^2 d\mathfrak{v} > 0. \tag{75·10}$$

It may be noted in passing that, since $\dot{u} = c\dot{T}$, one has from (75.5, 8) the differential equation of thermal conduction

$$\operatorname{div}(k \operatorname{grad} T) = c\dot{T}. \tag{75.11}$$

Equation (75.10) simply confirms in detail that the entropy of a state which arises by thermal conduction from some initial unnatural state is greater than that of the latter. However, the history of this entropy increase is now traced through, a possibility which the purely classical theory does not admit. One must, however, keep in mind that the validity of (75.10) is contingent upon the possibility of assigning local temperatures, as observed previously, and is therefore restricted to conditions which in some sense do not differ too widely from equilibrium. If K is not isotropic, the vectorial character of $\mathbf{q}$ and $\operatorname{grad} T$ ($\equiv T_{,1}, T_{,2}, T_{,3}$) implies that

$$\mathbf{q} = -\sum_j k_{ij} T_{,j}, \tag{75.12}$$

where k_{ij} is a tensor of valence 2. In place of (75.10) one then has

$$\dot{S} = \int T^{-2} \sum_{i,j} k_{ij} T_{,i} T_{,j} d\mathfrak{v}. \tag{75.13}$$

Since one may deal with any selected part of K^* just as well as with K^* as a whole (by the use of adiabatic partitions) it follows from this, on applying the Entropy Principle to an arbitrarily selected pair of unnatural states, that one must have

$$k_{ij} T_{,i} T_{,j} \geqslant 0, \tag{75.14}$$

i.e. the 'thermal conductivity tensor' must be positive definite.

(*b*) The equation (75.10) not only describes the history of the entropy increase of K^*, but it has every appearance of showing the *local* details of entropy production. Quite generally there is a rate of entropy production e per unit time and unit volume:

$$e = -T^{-2}\mathbf{q} \cdot \operatorname{grad} T. \tag{75.15}$$

The meaningfulness of this expression should not depend on whether K^* is adiabatically isolated or not. Reversing the previous argument one then has

$$\dot{S} = -\oint T^{-1}\mathbf{q} \cdot d\mathfrak{f} + \int e\, d\mathfrak{v}. \tag{75.16}$$

Hence if $\mathbf{q}$ does not vanish on $\mathfrak{F}$ an obvious interpretation of (75.16) is that the first term on the right represents the rate at which entropy is added to K^* from without, as a consequence of energy flowing into the system, whilst the second term represents the rate of creation of entropy within K^*. In other words, one adopts the interpretation of $\mathbf{q}$ as an *energy flux* and of $\mathbf{s} = \mathbf{q}/T$ as an *entropy flux* associated with $\mathbf{q}$, whilst e is the *local rate of creation of entropy*. Thus

$$\operatorname{div}\mathbf{s}+\dot{s} = e, \tag{75.17}$$

the term on the right characterizing the non-conservation of entropy. When (75.8) applies one has the interesting equation

$$\operatorname{div}\mathbf{s}+\dot{s} = k^{-1}|\mathbf{s}|^2. \tag{75.18}$$

The term 'interpretation' was used just now quite deliberately: it is better to say guardedly that $\mathbf{q}$ is *interpreted* as an energy flux than to say that it *is* an energy flux. An analogous remark applies to the entropy flux. First of all, recall that it was assumed that $\mathbf{q}$ vanishes when $\operatorname{grad} T$ does so. Strictly speaking all one knows is that $\operatorname{div}\mathbf{q}$ vanishes when $\operatorname{grad} T$ does so. In other words, $\mathbf{q}$ is determined only to within an additive term which takes the form of the curl of an arbitrary vector. Consider therefore the consequences of a 'gauge transformation'

$$\mathbf{q}\to\mathbf{q}+\operatorname{curl}\mathbf{a}, \tag{75.19}$$

$\mathbf{a}$ arbitrary. Then

$$\mathbf{s}\to\mathbf{s}+T^{-1}\operatorname{curl}\mathbf{a}, \quad e\to e+\operatorname{div}(T^{-1}\operatorname{curl}\mathbf{a}). \tag{75.20}$$

This means that although (75.17) is of course invariant under gauge-transformations, the energy flux, the entropy flux, and rate of creation of entropy are not. The physical meaning of these quantities is therefore in doubt, in the sense that one should perhaps consider them to be merely auxiliary functions which are convenient for purposes of calculation. In some cases it may be possible to lend more weight to the interpretation of $\mathbf{q}$; and the present situation is perhaps of this kind. For one could argue that by considering enclosures which consist of some 'non-conducting' material except over very small parts of their surfaces, the energy transport can be *localized*. This sort of argument is, however, likely to be successful only in special cases, and there is no doubt that one must be wary against assuming that energy and its transport can be properly localized in general. One sees this already in the electrostatic field. Again, the Poynting vector is only defined to within an additive curl, a fact one does well to keep in mind when confronted with the odd conclusion that this vector does not vanish in crossed electrostatic and magnetostatic fields. Of course the Poynting flux is physically meaningless in this case. In the gravitational field, finally, the absolute localization of energy becomes quite hopeless.

There is no obvious reason why what has been accomplished here for thermal conduction should not be generalized to processes of a more complex kind. The theory which achieves this is sometimes known as the 'thermodynamics of irreversible processes'. It contains a new law, called *Onsager's Principle*, which, though phenomenological in character, finds no place in the classical theory at all, in as far as it distinguishes quantitatively between different sources of irreversibility which may be present in a system, and governs the interference between them. To consider this extended theory would, however, lead altogether too far outside the compass of this book.

76. Fluctuations considered phenomenologically

(*a*) When two systems K_A and K_B, initially at different temperatures, are brought into mutual diathermic contact, the compound system K_C so formed will eventually attain a certain state such that the temperatures of K_A and K_B are then equal. Similarly, if two separate enclosures of equal volume which initially contain different amounts of a certain gas are joined together by a tube permitting the flow of matter, then eventually each enclosure will contain the same amount of gas, granted that the temperature is uniform throughout the compound enclosure. The possibility that, inversely, unnatural states of either system could spontaneously occur at some later time is excluded by the Second Law.

Now it is the case that if a gas, for instance, is left to itself for an indefinitely long time within a diathermic enclosure of given volume, then it is observed that (i) the energy of the gas as a whole will fluctuate about a mean value, and (ii) local density fluctuations will occur within the body of the gas. Operationally the system will therefore never be quiescent. For this reason classical thermodynamics has sometimes been held to be completely unrealistic, since the central concept of a state implies quiescence. This objection can surely be given short shrift. Whatever the operations may be that will yield a value of a pressure P, for example, one need only interpret P as the *mean* of the values obtained by carrying out generically identical measurements on the given system a large number of times. One may read a thermometer a large number of times at regular intervals and take the mean; and the mean value of any other quantity may be obtained in similar fashion. A *state* is then to be taken as a set of mean values and the development of the theory proceeds much as before, though one may encounter some verbal difficulties.

(*b*) If the fluctuations are to be described quantitatively, one is, as in the preceding section, confronted with the need to extend the classical theory. This being so, the problem will only be dealt with briefly. However, it deserves some space here on the grounds that it is amenable to phenomenological treatment. For the sake of orientation, consider a system K which is in diathermic contact with some much larger system K^e, the two together forming a compound system K^*. The latter shall be regarded as adiabatically isolated. The volume of each system is given and remains fixed. K^* being in equilibrium, the temperatures, in the sense of mean values, of K^e and K are equal, T being their common value. The possibility of fluctuations implies that the energy of K will not be constant. This means the following: imagine K to be surrounded instantaneously by an adiabatic envelope at time t. Thereafter its energy is fixed. Of course, one may wish to measure this by measuring its temperature. It might be argued that one then has to establish contact with a thermometer, which destroys the adiabatic isolation. This is true but irrelevant, provided (i) the temperature in question is understood to be a mean value, and (ii) the specific heat of the thermometer is negligible compared with that of K. In this way, then, one can assign a value to the 'energy of K at time t'. Let the whole procedure just described be repeated a large number of times. The question is: what is the relative frequency of occurrence $\phi(U)dU$ of a value of the energy of K which lies in the range $(U, U+dU)$?

In a more general situation one might be interested in the simultaneous fluctuations of several quantities, generically denoted by $\xi_1, \ldots, \xi_\alpha$. These will occur as a consequence of the coupling between K and K^e; and the values of $\xi_1, \ldots, \xi_\alpha$ will be the mean values measured *after* decoupling. It will be seen that, in general, after decoupling the state of K and the corresponding state of K^e together make up an unnatural state of K^*. If one recalls that a maximum of the entropy function, calculated for a range of unnatural states, corresponds to equilibrium, one will suspect that an unnatural state is the less likely the more the value S^* of its pseudo-entropy deviates from its maximum $\bar{S}^*$.

Write

$$\Delta\xi_\beta = \xi_\beta - \bar{\xi}_\beta, \quad \Delta S^* = S^* - \bar{S}^*, \tag{76.1}$$

bars denoting mean values. S^* is a function of the ξ_β, or more conveniently, of the $\Delta\xi_\beta$, for then ΔS^* vanishes when the $\Delta\xi_\beta$ do so. Without further ado the following *postulate* will now be adopted:

The relative frequency of occurrence of a state of K in which the quantities ξ_β have values in the range ξ_β, $\xi_\beta + d\xi_\beta$ $(\beta = 1, \ldots, \alpha)$ is

$$\phi(\xi_1, \ldots, \xi_\alpha)\, d\xi_1 \ldots d\xi_\alpha = A e^{\Delta S^*/k}\, d\xi_1 \ldots d\xi_\alpha, \tag{76.2}$$

where k is a universal constant, and A a normalizing constant.

It will hereafter be assumed that all fluctuations are small. Then the problem of the energy fluctuations already introduced above may be solved as follows. Take S^* as a function of the energy U of K, the volumes being left understood, since they are fixed. From (70.2) one has in the present situation

$$\Delta S^* = -(1/2\bar{T}^2)\,(1/\bar{c} + 1/\bar{c}^e)\,(\Delta U)^2, \tag{76.3}$$

where powers of ΔU higher than the second have been neglected, on account of the assumed smallness of the fluctuations. One is primarily interested in a situation where the fluctuations in K are not influenced explicitly by the character of K^e, and this requires that the latter be taken as sufficiently large. Then $c^e \gg c$, and, with $\eta = \Delta U$, (76.2, 3) reduce to

$$\phi = A \exp(-a^2\eta^2), \quad a^{-2} = 2kc\bar{T}^2. \tag{76.4}$$

Since ϕ is a relative frequency, $\int \phi\, d\eta$ must have the value unity, the lower and upper limits of integration corresponding to the least possible values of U and U^e respectively. However, the contributions to the integral from the regions where η^2 is large are negligible, and so the integration may be extended from $-\infty$ to $+\infty$. Then

$$A = \pi^{-\frac{1}{2}}a. \tag{76.5}$$

A suitable measure of the energy fluctuation is the root mean square deviation from the mean

$$\delta U = \overline{(\eta^2)}^{\frac{1}{2}}. \tag{76.6}$$

Thus $$(\delta U)^2 = A\int \eta^2 \exp(-a^2\eta^2)\,d\eta = \tfrac{1}{2}\sqrt{\pi}\,a^{-3}A,$$

whence $$\delta U = (k\bar{c})^{\frac{1}{2}}\bar{T}. \tag{76.7}$$

This result is well known from statistical mechanics, provided k be identified with Boltzmann's constant.

The elimination of the effects of the size of K^e was achieved by supposing it to be large. This means that a small fluctuation in K will not affect P^e and T^e. All unnatural states of K^* to be admitted for consideration therefore have $T^e = \bar{T}$, $P^e = \bar{P}$, so that

$$\Delta S^e = \int (dU^e + \bar{P}dV^e)/\bar{T} = -(\Delta U + \bar{P}\Delta V)/\bar{T}.$$

Hence ϕ may be put in a form which makes reference to K alone:

$$\phi = A\exp[(\bar{T}\Delta S - \Delta U - \bar{P}\Delta V)/k\bar{T}]. \tag{76.8}$$

If two quantities fluctuate simultaneously, write $\Delta\xi_i = \eta_i$ $(i = 1, 2)$. Then ϕ will have the generic form

$$\phi = A\exp[-(a_{11}\eta_1^2 + 2a_{12}\eta_1\eta_2 + a_{22}\eta_2^2)]. \tag{76.9}$$

Then

$$J \equiv A^{-1}\iint \phi\, d\eta_1\, d\eta_2 = \pi(a_{11}a_{22} - a_{12}^2)^{\frac{1}{2}} = \pi a^{-\frac{1}{2}}, \quad \text{say.} \tag{76.10}$$

The determinant a of the a_{ik} is certainly positive. By inspection

$$(\delta\eta_i)^2 = \overline{\eta_i^2} = -\partial \ln J/\partial a_{ii} \quad (i = 1, 2),$$

and $$\delta(\eta_1\eta_2) = \overline{\eta_1\eta_2} = \tfrac{1}{2}\partial \ln J/\partial a_{12},$$

so that

$$(\delta\eta_1)^2 = \tfrac{1}{2}a_{22}/a, \quad (\delta\eta_2)^2 = \tfrac{1}{2}a_{11}/a, \quad \delta(\eta_1\eta_2) = -\tfrac{1}{2}a_{12}/a. \tag{76.11}$$

If ξ_2 were kept fixed so that ξ_1 alone could fluctuate one would get

$$(\delta\eta_1)^2 = \tfrac{1}{2}a_{11}^{-1}, \tag{76.12}$$

and the two values of $\delta\eta_1$, corresponding to the two different physical situations, are equal only if a_{12} happens to be zero. This occurs, for instance, when one takes $\xi_1 = T$, $\xi_2 = V$. The first-order terms of the exponent must of course be zero, whilst the second-order terms are readily verified to be given by

$$a_{11} = \bar{c}/(2k\bar{T}^2), \quad a_{12} = 0, \quad a_{22} = -(2k\bar{T})^{-1}\overline{P_{,V}}.$$

The derivative $\overline{P_{,V}}$ refers to constant T, and is evaluated at $\bar{T}$, $\bar{V}$. Then

$$\delta T = (k/\bar{c})^{\frac{1}{2}}\bar{T}, \quad \delta(T, V) = 0, \quad \delta V = (-k\bar{T}\overline{V_{,P}})^{\frac{1}{2}}, \tag{76.13}$$

where $\delta(\xi_1, \xi_2)$ stands for $(\overline{\Delta\xi_1\Delta\xi_2})^{\frac{1}{2}}$. The first of these is entirely equivalent to (76.7) when V cannot fluctuate. When it can one gets, however, a different result. Thus, from

$$\Delta U = c\Delta T + U_{,V}\Delta V$$

it follows directly, after squaring and inserting the previous values (76.13) for $\overline{(\Delta T)^2}$ etc., that

$$(\delta U)^2 = kcT^2 - kT(U_{,V})^2 V_{,P}, \tag{76.14}$$

where bars have been dropped. For an ideal gas the second term on the right vanishes.

In deriving general relations of this kind one has to be careful to relate the details of the calculations to the actual physical situation in hand, or one runs the risk of getting discordant results. However, it would clearly be out of place here to pursue the extended theory any further.

CHAPTER 10

APPLICATIONS (I)

77. Classification of applications

As a matter of principle it is quite useful to classify the various applications—meaning the more detailed consequences—which arise from the basic theory developed in preceding chapters. In this way one gets a somewhat more orderly picture of the manner in which specific results follow from the general laws. To begin with there are two broad types of applications, to be denoted generically by the numerals 1 and 2: type 1 covers all situations in which only states are contemplated, whilst type 2 rests upon the basis of the Entropy Principle, that is to say, unnatural states are admitted. If one adopts the view-point of Section 64, all problems involving questions of physico-chemical equilibrium are evidently of type 2, whereas identities between thermodynamic functions and their derivatives are of type 1. With regard to the latter one has subclasses, distinguished by subscripts, such that an application is of type 1_0 if it is based on the Zeroth Law alone, of type 1_1 if it is based on the Zeroth and First Laws, and so on. In the case of type 2 one has the subclasses 2_2 and 2_3. An application of any particular kind may or may not involve the introduction of extraneous assumptions. The presence of such non-thermodynamic assumptions may be indicated by an additional asterisk. By way of example, the derivation of the equation of state (71.12) of a super-ideal gas is of type 1_2^*.

Applications of types 1 and 2 go to make up Chapters 10 and 11 respectively. At times the distinction between them is somewhat diffuse on account of the availability of the alternative viewpoints represented by Sections 63 and 64. Thus the equations connecting the equilibrium pressure and temperature of a one-constituent two-phase system (the Clapeyron–Clausius equation, Section 83(v)) may be taken to describe a state of affairs in which (stable) equilibrium is taken for granted: in that case it is of type 1_2, and it is here regarded as such.

Where appropriate, some particular equation, rather than the situation in which it occurs, will be said to be of a certain type; and the latter will at times simply be indicated in square brackets. Broadly speaking progress will be through the various types in turn. However, as previously emphasized, the choice and manner of presentation of particular examples is governed by the general orientation of this book, which is towards general principles rather than specific results.

78. Consequences of the existence of an equation of state

Suppose K is a standard system with n coordinates $x_1, \ldots, x_n$. If t is some empirical temperature, the equation of state of K may be written in the form

$$F(x_1, \ldots, x_n, x_{n+1}) = 0 \quad (x_{n+1} = t). \tag{78.1}$$

Granted that this equation is soluble for any one of the $n+1$ variables, x_j is a function f_j of all the remaining variables. Then the function $F(x_1, \ldots, x_{j-1}, f_j, x_{j+1}, \ldots, x_{n+1})$ vanishes *identically*. For all j, k $(j \neq k)$ one therefore has in the notation of Section 33*a*

$$F_{,k} + F_{,j} f_{j,k} = 0.$$

Choose $k = j+1$, with $j = 1, 2, \ldots, r-1$ in turn, and adjoin the equation corresponding to $j = r$, $k = 1$. Consistency of the resulting set of r equations requires that

$$f_{1,2} f_{2,3} \cdots f_{r-1,r} f_{r,1} = (-1)^r. \tag{78.2}$$

In the case of a simple homogeneous and isotropic substance write T for the empirical temperature (taking this to be the 'gas-temperature' for instance—recall (49.5–8)—until such time as the absolute temperature has been defined). Equation (78.2) then gives, with $n = 2$, $r = 3$,

$$V_{,T} T_{,P} P_{,V} = -1. \tag{78.3}$$

Since it is obvious from the context what variables are being kept constant in the differentiations, it is not necessary here to indicate them by further subscripts.

One usually defines certain phenomenological 'coefficients' as follows:

(i) the isobaric coefficient of expansion

$$\alpha = (\ln V)_{,T},$$

(ii) the isometric pressure coefficient

$$\beta = (\ln P)_{,T},$$

(iii) the isothermal compressibility

$$\kappa = -(\ln V)_{,P}. \tag{78.4}$$

In the case of a solid α and κ are easily measured, though β is not. However, from (78.3) one has [1_0]

$$\beta = \alpha / P\kappa. \tag{78.5}$$

If $n > 2$ one has a multiplicity of relations of this kind which can be written down. A paramagnetic gas for instance, has $n = 3$, and so there will be three identities of the kind (78.3), each involving three derivatives, and one which involves four derivatives.

Another kind of identity is obtained as follows. Using the definition of α and κ one has

$$dV/V = \alpha dT - \kappa dP,$$

the coordinates here being P and T. Then since dV/V is a total differential one must have [1_0]

$$\alpha_{,P} = -\kappa_{,T}. \tag{78.6}$$

79. Virial coefficients

(*a*) The equation of state of a specific substance has to be determined empirically. As previously remarked, many real gases obey (71.2, 3) quite closely under suitable conditions. Of course, the absolute temperature is not yet available. However, it is convenient to proceed here, and in the next section, as if it were. (See the remark following (79.1).) For practical purposes the imperfections of a nearly ideal gas, so far as these affect the equation of state, may be exhibited conveniently by the so-called virial expansion, i.e. the expansion of the *virial* $Z = PV/RT$ in descending integral powers of V. It is assumed that this expansion exists and converges. Experiment shows that if the imperfections are disregarded altogether (V and PV sufficiently large), PV/T has the same value R for one mole of every gas. Then

$$Z = 1 + BV^{-1} + CV^{-2} + \ldots, \tag{79.1}$$

where the second, third, ... *virial coefficients* B, C, ... are functions of T alone. (If T is not the absolute temperature the first term on the right is not unity but some function A of the temperature; and the functional forms of A, B, C, ... depend on the particular temperature scale which is being used.) As a first approximation it may suffice to suppose only B to differ from zero, and this coefficient may then be easily determined experimentally by studying effects whose magnitude depends on the derivatives of Z. In any event, whatever form of some equation of state one uses to begin with, it should be reduced to the form (79.1) before one makes any attempt at a quantitative interpretation of constants which appear in it in terms of particular molecular models.

The virial expansion (79.1) applies whether the gas in question is simple or a mixture. In the latter case the virial coefficients will be functions of the concentrations $c_1, c_2, \ldots$ of the constituents. Thus

$$B = \sum_{i,j} B_{ij} c_i c_j, \quad C = \sum_{i,j,k} C_{ijk} c_i c_j c_k, \tag{79.2}$$

and so on, the coefficients $B_{ij}, C_{ijk}, \ldots$ being symmetric in all their indices. Confining attention to B for the moment, B_{ii} is the same as the virial coefficient for one mole of the ith constituent taken by itself; but the B_{ij} $(i \neq j)$ are new coefficients. If one has n moles altogether the rth virial coefficient in (79.1) must be supplied with an additional factor n^r.

(*b*) The behaviour of real gases is so complex that it is a fruitless exercise to seek an equation of state in closed form, unless one admits a very large number of phenomenological constants into it. One may as well stick to an expansion of the kind (79.1), except in as far as a series in ascending powers of P may occasionally be more convenient. For illustrative purposes on the other hand some closed equation of state is useful at times. The one quoted most often is that of van der Waals:

$$(P + aV^{-2})(V - b) = RT, \tag{79.3}$$

where a and b are positive constants. (Even under conditions in which gases are nearly ideal, a is not in fact independent of T.) The term aV^{-2} is supposed to be a correction representing the attractive forces which the interior parts of the gas exert upon the boundary layer—hence the dependence on the square of the density—whilst the constant b ensures that the compressibility becomes zero at some positive volume. For the van der Waals gas

$$B = b - a/RT, \tag{79.4}$$

whilst for $r > 2$ the rth virial coefficient is simply b^{r-1}. Equation (79.3) will be used later on various occasions.

With regard to the super-ideal gas one has, according to (71.12), $Z = g(z)$. Consistency with (79.1) requires that the mth virial coefficient $Z^{(m)}$ have the form

$$Z^{(m)} = z^{(m)} T^{-(m-1)/r}, \tag{79.5}$$

where $z^{(m)}$ is a constant, $z^{(1)} = 1$.

For further remarks concerning equations of state reference may be made to Section 85.

80. Consequences of the existence of the energy function

The scope of the inferences which may be drawn by invoking the existence of U is rather narrow, largely because until the existence of S is invoked as well the form of the energy function of a system is in no way restricted by its equation of state. Nevertheless it is instructive to consider very briefly what can be done.

For an isotropic homogeneous substance, usually a fluid, one has, in the absence of electromagnetic effects,

$$dQ = dU + PdV,$$

or, if V and T are taken as coordinates,

$$dQ = U_{,T}dT + (U_{,V} + P)dV. \tag{80.1}$$

Now the specific heat C for a given process is defined as

$$C_* = (dQ/dT)_*, \tag{80.2}$$

where the asterisk is to be understood as symbolizing the statement of those conditions which make the process well defined. For example, if it takes place at constant volume the asterisk will become a (subscript) V, so that C_V is just what has hitherto been denoted by c. A derivative such as $U_{,P}$, written in this way, is in general ambiguous. Unless it is obvious from the context as to which are the variables of which U is regarded as a function, one has to adopt one of the three alternatives set out towards the end of Section 12*a*. The third of them is the conventional choice; it is convenient, and where the avoidance of ambiguity demands it, it will be used here, except that the 'comma-subscript' notation for derivatives will be retained. For example, the partial derivative of $U(P, T)$ with respect to T will be written $(U_{,T})_P$. (If U were a function of three coordinates, say P, T and I, one would have to write $(U_{,T})_{PI}$.) Now

$$C_V = (U_{,T})_V, \quad C_P = (U_{,T})_V + [(U_{,V})_T + P]\,(V_{,T})_P.$$

Hence [1_1]
$$C_P - C_V = [(U_{,V})_T + P]\,(V_{,T})_P. \tag{80.3}$$

If the equation of state is $PV = RT$ one has [1_1^*]

$$C_P - C_V = R + RP^{-1}(U_{,V})_T. \tag{80.4}$$

The constancy of $C_P - C_V$ does not follow yet. For an ideal gas, on the other hand, the last term of (80.4) vanishes by definition. Again, if P and V be used as coordinates,

$$U_{,V} = C_P T_{,V} - P, \quad U_{,P} = C_V T_{,P},$$

whence
$$(C_P T_{,V})_{,P} - (C_V T_{,P})_{,V} = 1. \tag{80.5}$$

If $PV = RT$ it follows that [1_1^*]

$$C_P - C_V = R + (VC_{V,V} - PC_{P,P}). \tag{80.6}$$

As a further application of type 1_1 one might consider the processes of Joule expansion and Joule–Thomson expansion. However, it is more sensible to defer doing so until the existence of S has been drawn into the argument, and reference may be made to Sections 83(i) and (ii).

81. Consequences of the existence of *S*. Thermodynamic identities

(*a*) Once the entropy S and absolute temperature T are available the range of possible relations of interest which might be written down is vastly enhanced. This is to some extent due to the fact that once an equation of state is known the energy function is already determined to within an unknown function of T only.

Consequently, the properties of systems, that is, the description of quasi-static transitions which they may undergo, is to a large extent governed by the equation of state alone. Any such description will involve derivatives of one sort or another—(80.3) may serve as an example—which make their appearance naturally, as it were, often merely as a matter of definition. These derivatives are then as far as possible to be replaced by others which can be evaluated on the basis of the equation of state. The mutual equality of a pair of derivatives is in every case the expression of a condition of integrability.

Perhaps the most important example here is the following. One has generally

$$\begin{aligned} dS &= T^{-1}(dU+\Sigma P_i\,dx_i) \\ &= T^{-1}[U_{,T}\,dT+\Sigma(U_{,i}+P_i)\,dx_i]. \end{aligned} \tag{81.1}$$

T is now the absolute temperature, and, as in Section 33, a subscript i following a comma denotes differentiation with respect to x_i. The coordinates are of course $x_i, \ldots, x_{n-1}, T$. Since the right-hand member of (81.1) is a total differential one has the integrability conditions

$$(T^{-1}U_{,T})_{,i} = [T^{-1}(U_{,i}+P_i)]_{,T},$$

i.e.

$$U_{,i} = TP_{i,T}-P_i, \tag{81.2}$$

of which (59.3) is a special case. (One has of course also the remaining conditions $P_{i,k} = P_{k,i}$.) If one has only one deformation coordinate V, (81.2) reads [1[2]]

$$U_{,V} = TP_{,T}-P. \tag{81.3}$$

Thus, when the equation of state, say in the form $P = P(V, T)$, is known, U is determined to within an unknown additive function $U^\circ(T)$ of T. Explicitly [1[2]],

$$U = U^\circ(T)+T^2\int(P/T)_{,T}\,dV. \tag{81.4}$$

A similar conclusion holds in the case of several deformation coordinates, so that one has here the connection between energy function and equation of state which was referred to above.

From (81.3) it now follows that a gas whose equation of state is $PV = RT$ is automatically ideal, since the right-hand member of (81.3) then vanishes.

(*b*) The four functions U, F, G, H all have the dimensions of energy, and they are therefore conveniently considered together, with a view to obtaining a set of integrability conditions. To simplify the notation, only one deformation coordinate V will be supposed to be involved. As explained in Section 66*a*, the equations appropriate to a more general case are easily written down. From (62.4–7),

$$\left.\begin{aligned} dU &= TdS - PdV, \quad & dF &= -SdT - PdV, \\ dG &= -SdT + VdP, \quad & dH &= TdS + VdP. \end{aligned}\right\} \tag{81.5}$$

Then, at once [1_2],

$$\left.\begin{aligned} P_{,S} &= -T_{,V}, \quad & S_{,V} &= P_{,T}, \\ S_{,P} &= -V_{,T}, \quad & V_{,S} &= T_{,P}. \end{aligned}\right\} \tag{81.6}$$

The subscripts indicating the variables to be held constant have been omitted here, since the two relevant independent variables in each case already appear as subscripts. However, they may of course be included in order to prevent the occurrence of errors when these identities are used elsewhere. The four relations (81.6) are known as *Maxwell's relations*. In place of (81.5) one may also consider the differentials of the modified potentials $\tilde{F}$ and $\tilde{G}$. The first merely gives (81.3) again (cf. (62.14)), and this is a good illustration of the remark that every such identity is simply an integrability condition. As regards $d\tilde{G}$ one gets, if the coordinates are P and T,

$$H_{,P} = -TV_{,T} + V, \tag{81.7}$$

which evidently corresponds closely to (81.3). One can increase the number of identities endlessly, for example by merely rewriting the relations (81.5) to get the integrability conditions on $dV, dP, \ldots$ with various choices of independent variables. However, to do so is hardly more than a somewhat empty exercise, since the occurrence of such rather esoteric derivatives in a calculation merely means that this is being carried out in a clumsy way.

Identities involving higher derivatives are of some interest at times. Thus, with V, T as coordinates, $C_{V,V} = U_{,TV}$, and using (81.3) on the right,

$$C_{V,V} = TP_{,TT}. \tag{81.8}$$

Similarly, with P, T as coordinates, since $C_P = H_{,T}$, there comes, on using (81.7),

$$C_{P,P} = -TV_{,TT}. \tag{81.9}$$

The coordinates which are being used in each case must be firmly kept in mind: it would be a mistake to insert (81.8, 9) directly into the right-hand member of (80.6), since the coordinates there were P and V.

(*c*) Previous relations of type 1_1 may now be strengthened, as shown in the example of (80.3). Using (81.3) this becomes [1_2]

$$C_P - C_V = T(V_{,T})_P (P_{,T})_V. \qquad (81.10)$$

In practice the presence of the derivative $(P_{,T})_V$ is inconvenient. It may be eliminated by means of (78.3), giving

$$C_P - C_V = -T(P_{,V})_T (V_{,T})_P^2. \qquad (81.11)$$

In terms of the bulk coefficients (78.4) this reads

$$C_P - C_V = TV\alpha^2/\kappa \geqslant 0. \qquad (81.12)$$

If the equation of state is given explicitly in the form $P = P(V, T)$ it is better to take

$$C_P - C_V = -T(P_{,T})^2/P_{,V}. \qquad (81.13)$$

For a van der Waals gas (79.3) gives

$$C_P - C_V = R[1 - 2a(V-b)^2/RTV^3]^{-1}. \qquad (81.14)$$

In particular, for an ideal gas

$$C_P - C_V = R. \qquad (81.15)$$

(*d*) The various thermodynamic identities considered hitherto in this section involved only the external coordinates. This restriction may be removed. Thus from (66.7), i.e.

$$dG = V\,dP - S\,dT + \sum_i \mu_i dn_i, \qquad (81.16)$$

one has further integrability conditions, viz.

$$\mu_{i,n_k} = \mu_{k,n_i}, \qquad (81.17)$$

$$\mu_{i,P} = V_{,n_i}, \quad \mu_{i,T} = -S_{,n_i}. \qquad (81.18)$$

From (66.8) one has

$$dG = \sum_i (n_i d\mu_i + \mu_i dn_i),$$

which may be combined with (81.16) to give the *Gibbs–Duhem identity*

$$S\,dT - V\,dP + \sum_i n_i d\mu_i = 0, \qquad (81.19)$$

so that the intensive variables $P, T, \mu_1, \ldots, \mu_z$ cannot change independently of one another in a quasi-static transition. For each phase of a heterogeneous system one has one such identity. Thus

$$S^k dT - V^k dP + \sum_i n_i^k d\mu_i^k = 0 \quad (k = 1, \ldots, p). \qquad (81.20)$$

82. Thermodynamic functions of gases and gas mixtures

For illustrative purposes and for later applications alike it is convenient to have various thermodynamic functions for classes of gases and gas mixtures available once and for all. This section therefore concerns itself with collecting relevant results. These will be written in a way designed to emphasize their phenomenological character, in the sense exemplified by the equation

$$U = \int_{T_1}^{T} C_V(T)\, dT + U_1 \qquad (82.1)$$

for the energy of an ideal gas. U_1 is a constant whilst C_V is accessible to direct (calorimetric) measurement. T_1 is some standard temperature, which must be taken sufficiently large to ensure that the operation of the Third Law does not come in conflict with the assumed equation of state.

(i) *Single ideal gas.* For one mole, by definition,

$$PV = RT, \qquad (82.2)$$

whilst U is given by (82.1), and is a function of T only. Then

$$S(V, T) = \int_{T_1}^{T} \tilde{C}_V\, dT + R \ln V + \text{const.}, \qquad (82.3)$$

where $\tilde{C}_V = C_V/T$ as usual. If P, T be taken as coordinates

$$S(P, T) = \int_{T_1}^{T} \tilde{C}_P\, dT - R \ln P + S_1, \qquad (82.4)$$

where S_1 is a constant. Then

$$G(P, T) = RT \ln P + \int_{T_1}^{T} C_P\, dT - T \int_{T_1}^{T} \tilde{C}_P\, dT - S_1 T + U_1. \qquad (82.5)$$

If one wishes to emphasize that (82.5) represents the *molar* Gibbs function one may use lower-case symbols:

$$g(P, T) = RT \ln P + \gamma(T) - s_1 T + u_1. \tag{82.6}$$

Here
$$\gamma(T) = \int_{T_1}^{T} c_P dT - T \int_{T_1}^{T} \tilde{c}_P dT, \tag{82.7}$$

c_P being the molar specific heat. Note that

$$c_P = -T\ddot{\gamma}, \tag{82.8}$$

dots denoting differentiation with respect to T. If one does not like the appearance of dimensionally nonsensical terms such as $RT \ln P$ one may write $RT \ln(P/P^\dagger)$ instead, where $P^\dagger$ is some standard pressure, and interpret the constants of integration accordingly.

(ii) *Mixture of ideal gases.* An adiabatic enclosure is divided into two parts by means of a rigid diathermic partition. The first compartment (volume V_1) contains n_1 moles of a gas at pressure P and temperature T, whilst the second (volume V_2) contains a different gas at the same pressure and of course the same temperature. The partition is withdrawn and mutual diffusion sets in. Experiment shows that when the gases are sufficiently nearly ideal the state eventually attained by the mixture has the same pressure and temperature as each gas had initially, the absence of chemical reactions being presupposed. In fact, one may strengthen the *definition* of an ideal gas by requiring that this situation should obtain, in which case experiment shows that under suitable conditions the behaviour of real gases will in fact closely approximate that of ideal gases. Thus, the pressure of the mixture is the sum of the pressures which each gas would exert if it were present by itself. The separate pressures so defined are called *partial pressures.* Generally, for a mixture of z ideal gases, the partial pressure of the ith is $P_i = n_i RT/V$. But $RT/V = \bar{n}R$ (cf. (66.10–12)) so that

$$P_i = c_i P. \tag{82.9}$$

Now let K_C be a compound system, made up of two subsystems K_A and K_B, the latter being enclosures of volumes V_A, V_B respectively. K_A and K_B are connected by a tube which incorporates a semipermeable membrane which permits the free passage of a certain ideal gas C_1, but that of another ideal gas T_2 not at all. (It is a necessary hypothesis that such a membrane is always realizable in principle, though to find one in practice may be an awkward proposition, to say the least.) Initially K_A shall contain a mixture of n_1 moles of C_1 and n_2 moles of C_2, the volume V_A then being V, whilst V_B is zero. Further, K_C is in diathermic interaction with its surroundings so that its temperature is T at all times. Let V_B now be increased quasi-statically until it is so large that V_A is negligibly small compared with it. Then, since C_2 can pass freely between K_A and K_B, the latter will effectively contain all of C_2. Now replace the semipermeable membrane by one which is imperme-able, and reduce the volume V_B until it has the volume V. The temperature everywhere being T at all times, the pressure within K_B depends solely on the total volume available to C_2, and this was V both initially and finally. It follows that the work done by K_C in increasing the volume of K_B from zero was equal to the work done on K_C in subsequently reducing it to V. Hence $W_C = 0$, and, since the energy of ideal gases depends on T only, $\Delta U_C = 0$. The entire process having been quasi-static, it follows that $\Delta S_C = 0$. Thus, *the entropy of a mixture of ideal gases is the sum of the entropies which each of the constituents would have if, at the same temperature, it occupied the whole volume of the mixture.*

This result clearly holds for a mixture of any number of ideal gases. If it consists of n_i moles of C_i $(i = 1, \ldots, z)$ then the con-tribution to the entropy by one mole of C_i is given by (82.4). P is taken now as the *partial* pressure P_i, since this is the pressure which it would exert if it occupied the whole volume of the mixture. Using (82.9), and summing over all the constituents, the entropy of the mixture is given by

$$S(P, T) = \sum_{i=1}^{z} n_i \left(\int_{T_1}^{T} \tilde{c}_{Pi} dT - R \ln P - R \ln c_i + s_{1i} \right). \tag{82.10}$$

Thus
$$S(P, T) = \sum_{i} n_i (s_i - R \ln c_i), \tag{82.11}$$

where s_i is the molar entropy of C_i at pressure P and temperature T.

The following point of interest may be noted in passing. If each constituent gas were contained in a separate enclosure, all at the same pressure P and temperature T, their total entropy would be

$\sum_i n_i s_i$. If the walls between the separate enclosures be now removed, mere diffusion will set in, since the pressure is already homogeneous throughout. The final entropy is given by (82.11), which shows that the irreversible process of diffusion is accompanied by an increase of entropy ('entropy of mixing')

$$\Delta S = -R\sum_i n_i \ln c_i. \tag{82.12}$$

(This is certainly positive since $c_i < 1$.) This result is independent of pressure and temperature, and also of the nature of the (ideal) gases, except to the extent that the gases must be physically *distinct* in one way or another. In particular, if one has mutual diffusion of one mole of each of two gases,

$$\Delta S = 2R\ln 2. \tag{82.13}$$

If the two gases are identical there is no sense in talking about their mutual diffusion, and the result $\Delta S = 0$ is therefore not inconsistent with (82.13).

Finally, the Gibbs function of the mixture is

$$G(P, T, n_1, \ldots, n_z) = \sum_{i=1}^{z} n_i(g_i + RT\ln c_i), \tag{82.14}$$

where g_i is the molar Gibbs function of C_i at pressure P and temperature T, i.e. from (82.6),

$$g_i(P, T) = RT\ln P + \gamma_i(T) - s_{1i}T + u_{1i}. \tag{82.15}$$

The chemical potential of C_i *in the mixture* is evidently

$$\mu_i = g_i + RT\ln c_i. \tag{82.16}$$

This indeed depends on the mole numbers only through the concentrations, in harmony with the general result stated at the end of Section 66*c*.

(iii) *Single real gas.* All results will be derived under the assumption that effects of the third and higher virial coefficients are entirely negligible, which requires that the density should be sufficiently low. Thus, for n moles, one has the equation of state

$$P = nRT/V + n^2RTB(T)/V^2, \tag{82.17}$$

or equivalently
$$V = nRT/P + nB(T). \tag{82.18}$$

Using (81.4) as usual, one gets

$$U = U^\circ(T) - n^2RT^2\dot{B}/V. \tag{82.19}$$

(Note that now $U^\circ(T) \neq \int C_V dT$.) It is convenient to introduce the following notation. If X is any thermodynamic function, X^* shall be the same function calculated *as if* all virial coefficients after the first were zero. In the case of (82.19) this simply means replacing U° by U^*; and then $U^* = \int C_V^* dT$. The starred quantities are thus in effect the thermodynamic functions of an ideal gas. Proceeding in the usual way, one gets

$$S(V, T) = S^*(V, T) - n^2R(T\dot{B} + B)/V, \tag{82.20}$$

and

$$G(P, T) = G^*(P, T) + nBP. \tag{82.21}$$

Such results may be obtained by a short-cut, as the example of G will show. Given $V = V(P, T)$ one has from the first member of (66.6)

$$(G - G^*)_{,P} = V - V^*,$$

whence, integrating at constant T,

$$G = G^* + \int_0^P (V - V^*)\,dP, \tag{82.22}$$

where the arbitrary additive function of T only which should in principle appear on the right has been omitted since $G - G^*$ goes to zero with P. Using (82.18) in (82.22) one immediately recovers (82.21).

(iv) *Mixture of real gases.* The total number Σn_k of moles in a mixture has hitherto been denoted by $\bar{n}$. In future the bar will be omitted. Though the number of external coordinates of a general system will continue to be denoted generically by n, confusion between the alternative meanings attached to this symbol is most unlikely to arise. From Section 79*a* the equation of state is, to the required order,

$$PV = nRT + RTV^{-1} \sum_{i,j} n_i n_j B_{ij}(T), \tag{82.23}$$

or equivalently

$$V = n(RT/P + \sum_{i,j} c_i c_j B_{ij}). \tag{82.24}$$

Then

$$U = U^* - RV^{-1}T^2 \sum_{i,j} n_i n_j \dot{B}_{ij}. \tag{82.25}$$

The characteristic function $G(P, T, n_1, ..., n_z)$ is most easily obtained from (82.22). Using (82.24) one has straight away

$$G = G^* + nP \sum_{i,j} c_i c_j B_{ij}. \tag{82.26}$$

Hence one gets $S(P, T, n_1, ..., n_z)$ from the second member of (66.6):

$$S = S^* - nP \sum_{i,j} c_i c_j \dot{B}_{ij}. \tag{82.27}$$

The chemical potential of the ith constituent is

$$\mu_i = G_{,n_i} = g_i^* + RT \ln c_i + P(2 \sum_j B_{ij} c_j - \sum_{j,k} B_{jk} c_j c_k). \tag{82.28}$$

(v) *Van der Waals gas.* The equation of state for n moles is

$$(P + n^2 a V^{-2})(V - nb) = nRT, \tag{82.29}$$

where a and b are constants, i.e. independent of P, T, n. The energy function then has the form

$$U = U^\circ(T) - n^2 a/V, \tag{82.30}$$

so that the specific heat C_V depends on T only. Next

$$S(V, T) = n\left(\int_{T_1}^{T} \tilde{c}_V \, dT + R \ln(V/n - b) + \text{const.}\right). \tag{82.31}$$

The Gibbs function $G(V, T)$ then follows as usual. To get the characteristic function $G(P, T)$ from this one has to express V as a function of P and T by means of (82.29), and this unfortunately entails the solution of a cubic equation.

83. Miscellaneous results of Type I_2.

(i) *Joule expansion.* Two enclosures K_1 and K_2 of *fixed* volumes V and ΔV respectively form an adiabatically enclosed compound system K. Initially K_1 contains a gas C at temperature T whilst K_2 is vacuous. A hole is then made in the partition separating K_1 and K_2, so that C can flow from K_1 to K_2. As far as K as a whole is concerned this process of *Joule expansion* is evidently adiabatic and takes place without any work being done: $Q = W = 0$. Thus $\Delta U = 0$, and, if T was the initial temperature of C and $T + \Delta T$ its final temperature, ΔT is given by [I_1]

$$U(T + \Delta T, V + \Delta V) - U(T, V) = 0. \tag{83.1}$$

For an ideal gas ΔT vanishes of course. For one mole of a van der Waals gas one gets, in view of (82.30),

$$\Delta U^{\circ} = -a\Delta V/[V(V+\Delta V)]. \tag{83.2}$$

ΔT will be small in any event, so that (83.2) becomes

$$\Delta T = -a\Delta V/[c_V V(V+\Delta V)]. \tag{83.3}$$

On the other hand, if ΔV is sufficiently small (83.1) gives, on using (81.3),

$$\Delta T = -[(TP_{,T}-P)/c_V]\Delta V, \tag{83.4}$$

whatever the equation of state may be. When (82.17) holds, this yields, with $n = 1$,

$$\Delta T = -RT^2V^{-2}\dot{B}/c_V. \tag{83.5}$$

(ii) *Joule–Thomson expansion.* An adiabatically enclosed system K consists of n moles of a gas contained in two large enclosures A and B connected together by a tube containing an obstruction which is such that though the gas can flow evenly from A to B it cannot do so freely. Suppose that the volume of A is steadily decreased and that of B increased at a finite rate in such a way that the conditions within A and B are steady, i.e. the pressure and temperature P, T in A, and the pressure and temperature $P+\Delta P$, $T+\Delta T$ in B, do not vary in time. ΔP shall be so small that the kinetic energy of mass motion of the gas in A and B can be disregarded. Suppose then that under these conditions one mole of the gas disappears from A and reappears within B (*Joule–Thomson expansion*). To achieve this, the work done by K is

$$W = \int(P_A dV_A + P_B dV_B) = -P_A V_A + P_B V_B, \tag{83.6}$$

since the pressures are constant. On the right V_A and V_B are the *molar* volumes. Because of the adiabatic isolation $Q = \Delta U + W = 0$, so that in view of (83.6)

$$\Delta(U+PV) = \Delta H = 0, \tag{83.7}$$

where ΔX denotes the difference between the values of X calculated for one mole in A and B respectively. Since ΔP is small, ΔT will be small, and so

$$H_{,T}\Delta T = -H_{,P}\Delta P, \tag{83.8}$$

P and T being used as coordinates. Now $H_{,T} = c_P$, whilst $H_{,P}$ is given by (81.7). Hence

$$\Delta T = c_P^{-1}(TV_{,T}-V)\Delta P. \tag{83.9}$$

Depending on the actual conditions, ΔT may be positive or negative for given ΔP, and it may vanish at a certain temperature T_J, the *inversion temperature*. When (82.18) applies, (83.9) gives

$$\Delta T = c_P^{-1}(T\dot{B}-B)\Delta P. \tag{83.10}$$

Inserting here the form (79.4) of B appropriate to a van der Waals gas, one gets

$$\Delta T = c_P^{-1}(-b+2a/RT)\Delta P. \tag{83.11}$$

Accordingly

$$T_J = 2a/Rb. \tag{83.12}$$

At this temperature a van der Waals gas behaves like an ideal gas with respect to Joule–Thomson expansion. By contrast, a slightly imperfect van der Waals gas has the isothermal behaviour of an ideal gas at the *Boyle temperature*

$$T_B = a/Rb, \tag{83.13}$$

since B is then zero.

In (83.8) powers of ΔT and ΔP higher than the first have been ignored, whilst at the same time only the second virial coefficient has been retained in deriving (83.12). Now, according to (82.29, 30),

$$H = U^\circ(T)-2a/V+RTV/(V-b). \tag{83.14}$$

If this holds exactly for some gas, and if conditions are such that the kinetic energy of the gas can be disregarded, T will be given exactly by the conditions $\Delta H = 0$, $\Delta T = 0$, and one thus finds that

$$T_J = \frac{2a}{Rb}\left(1-\frac{b}{V}\right)\left(1-\frac{b}{V+\Delta V}\right). \tag{83.15}$$

Under normal experimental conditions (83.12) is therefore entirely adequate.

The Joule–Thomson expansion is of course irreversible, and there is a continuous creation of entropy at the approximate rate $RP^{-1}|\Delta P|$ per mole transferred.

(iii) *Remark on the mixing of real gases.* Suppose z chemically inert gases $C_1, C_2, \ldots$ are distributed over z separate compartments of a diathermal enclosure, the surroundings of which are at a fixed temperature T. The ith compartment contains n_i moles of C_i, so that, in view of (82.18), its actual volume is

$$V_i = n_i(RT/P+B_{ii}),$$

it being given that the pressure of every gas has the same value P.

The total volume of the enclosure is

$$V = n\left(RT/P + \sum_i c_i B_{ii}\right). \tag{83.16}$$

Now let all internal partitions be withdrawn. The total volume of the enclosure V is kept fixed. If $P+\Delta P$ is the value of the pressure of the mixture when equilibrium has set in, (82.23) shows that

$$P+\Delta P = nRT/V + RTV^{-2} \sum_{i,j} n_i n_j B_{ij}, \tag{83.17}$$

where V is to be inserted from (83.16). Doing so, one gets, to the required order,

$$\Delta P = (P^2/RT)\left(\sum_{i,j} c_i c_j B_{ij} - \sum_i c_i B_{ii}\right),$$

which may be written in the more attractive form

$$\Delta P = (P^2/2RT) \sum_{i,j} c_i c_j (B_{ii} - 2B_{ij} + B_{jj}). \tag{83.18}$$

The definition of ΔP should be carefully noted. Other quantities such as the heat of mixing under specified conditions, may be calculated in a similar fashion. With regard to the entropy of mixing, there will now be small correction terms to be added to the expression on the right of (82.12), but these are usually negligible.

(iv) *The Grüneisen parameter.* A quantity which turns up in certain simple theories of the solid state is the so-called *Grüneisen parameter*

$$\Gamma = \frac{\alpha V}{\kappa C_V}. \tag{83.19}$$

A suitable problem here is the investigation of the assumption that Γ is independent of temperature. The motivation is *Grüneisen's Law*, which does indeed state, on empirical grounds, that for certain classes of solids Γ does not depend on T. Now, in view of (78.5),

$$VP\beta = \Gamma(V) C_V.$$

By definition $C_V = T(S_{,T})_V$ and $P\beta = (P_{,T})_V$. Using the second of Maxwell's relations (81.6), there comes

$$VS_{,V} = \Gamma(V)\, TS_{,T}. \tag{83.20}$$

This is a simple differential equation, the solution of which is [1$_2^*$]

$$S = \psi(Tq), \tag{83.21}$$

where ψ is an arbitrary function of its argument Tq $(=y$, say), and

$$q = \exp\int \Gamma(V)dV/V. \tag{83.22}$$

When $\Gamma(V)$ = const., $q = V^\Gamma$, and then the isentropics of a substance obeying Grüneisen's Law are those of a super-ideal gas with $r = \Gamma$. P and V may now be found in the usual way. They are [1$_2^*$]

$$U = q^{-1}\int y\psi'(y)dy + \eta(V), \tag{83.23}$$

$$PV = q^{-1}\Gamma(V)\int y\psi'(y)dy - V\eta'(V), \tag{83.24}$$

where η is an arbitrary function and primes denote differentiation with respect to the argument. Note that

$$PV = \Gamma(V)U + \theta(V), \tag{83.25}$$

where $\theta = -[V\eta'(V) + \Gamma(V)\eta(V)]$.

(v) *First-order phase transitions.* Let two phases of a substance be in equilibrium. If this be taken for granted, the description of this equilibrium may be considered as an application of type 1_2, as already remarked in Section 77. According to (68.10) the chemical potentials of the substance in the two phases must be equal. Since the chemical potential of a substance, present by itself, is its molar Gibbs function, one therefore has

$$\Delta g = g^1(P, T) - g^2(P, T) = 0. \tag{83.26}$$

The equilibrium temperature is therefore determined by the pressure. The symbol Δ will be used, as in (83.26), to denote the difference between the molar quantities of the two phases. If $P+dP$, $T+dT$ is a neighbouring state of the two-phase system

$$d\Delta g = 0 = \Delta g_{,P}dP + \Delta g_{,T}dT. \tag{83.27}$$

Recalling (62.18), this gives, if v is the molar volume,

$$\frac{dP}{dT} = \frac{\Delta s}{\Delta v}. \tag{83.28}$$

The passage of one mole from the second to the first phase at pressure P and temperature T can occur reversibly, and is accompanied by the absorption of an amount of heat $\lambda = T\Delta S$, the *latent heat* of the phase change in question. One thus gets [1_2]

$$\frac{dP}{dT} = \frac{\lambda}{T\Delta v}, \tag{83.29}$$

which is the *Clapeyron–Clausius equation* for first-order phase transitions.

Equation (83.26) defines a curve in a representative space R_2, coordinates P, T. If one considers a third phase of the substance, this can coexist with the first two at one point of R_2 at most, called the *triple point*, which is the common point of intersection of the three curves defined by the equations

$$g^1 = g^2 = g^3 \tag{83.30}$$

taken in pairs.

The difference between the (molar) specific heats c_P^1 and c_P^2 is related to the temperature variation of the latent heat. Thus, differentiating λ/T with respect to T along the equilibrium curve, so that P is related to T through (83.29),

$$d(\lambda/T)dT = d(\Delta s)/dT = (\Delta s)_{,T} + (\Delta s/\Delta v)\,(\Delta s)_{,P}.$$

But $(\Delta s)_{,T} = (\Delta c_P)/T$ and $(\Delta s)_{,P} = -(\Delta v)_{,T}$, so that one gets *Clausius's equation* [1_2]

$$\Delta c_P = Td(\lambda/T)/dT + \lambda(\ln \Delta v)_{,T}. \tag{83.31}$$

First-order transitions in systems of several constituents will be considered in the next chapter.

(vi) *Higher-order phase transitions.* (*a*) It is characteristic of phase transitions of the first order that the Gibbs function of one mole of the substance has discontinuous first derivatives at a transition point, though G itself is continuous. This is just a statement of the physical situation. Now, proceeding somewhat formally, one may contemplate the possibility of G and its first, second, ..., $(m-1)$th derivatives being continuous, whilst its mth derivative is discontinuous. In such a case one speaks of an *mth-order transition.* This is just a matter of definition, and unexceptionable as such, unless G is not differentiable at all at the transition point. The identification of the character of a particular type of transition as it actually occurs in nature is a more difficult problem altogether.

The physical distinction between different phases becomes increasingly hazy with larger and larger m. At any rate, it will suffice to consider the case $m = 2$. Equation (83.28) is, of course, nugatory. On the other hand, though Δs and Δv vanish, the values of the derivatives of s differ in the two phases, and those of v likewise. In other words, one now has two equations corresponding to (83.27):

$$d\Delta s = 0 = \Delta s_{,P} dP + \Delta s_{,T} dT$$

and

$$d\Delta v = 0 = \Delta v_{,P} dP + \Delta v_{,T} dT.$$

Since $s_{,P} = -v_{,T} = -v\alpha$, $s_{,T} = c_P/T$, $v_{,P} = -v\kappa$, one thus has

$$\frac{dP}{dT} = \frac{\Delta c_P}{Tv\Delta\alpha}, \quad \frac{dP}{dT} = \frac{\Delta\alpha}{\Delta\kappa} \tag{83.32}$$

along the equilibrium curve (*Ehrenfest's equations*). It follows incidentally that [12]

$$\Delta c_P = Tv(\Delta\alpha)^2/\Delta\kappa. \tag{83.33}$$

It should be carefully noted that the derivation of (83.32) implied the assumption (which is not necessarily satisfied) that there is in fact an equilibrium *curve* representing second-order transitions along which one can differentiate.

(*b*) Physical examples of second-order transitions are not easily come by. One has to look for a situation in which at given pressure a substance transforms itself in its entirety into a different modification at some temperature, the transformation being unaccompanied by any latent heat or by any change of density. The best authenticated example of this kind of behaviour is provided, under appropriate conditions, by the change from the normal to the superconducting state of pure, crystalline superconducting elements. These states shall be distinguished by the superscripts 1 and 2 respectively. Consider a long thin cylinder of the metal, placed longitudinally in a uniform external magnetic field $\mathscr{H}$. $\mathscr{M}$ shall be, as before, the magnetic moment in the direction of $\mathscr{H}$, so that $-\mathscr{M}$ is a deformation coordinate and $\mathscr{H}$ the corresponding force. If one disregards all changes of volume of the cylinder (so that magnetostrictive effects are assumed absent), the Gibbs function of the system is

$$G = U - TS - \mathscr{H}\mathscr{M}. \tag{83.34}$$

(See, however, Section 83 (viii, *a*).) It is found experimentally that as $\mathscr{H}$ is increased the metal reverts from the superconducting to the normal state at a certain value $\mathscr{H}^*$ of the field strength, with $\mathscr{H}^*$ depending upon T. Whilst $\mathscr{H} < \mathscr{H}^*$ the magnetic induction $\mathscr{B} = \mathscr{H}+4\pi\mathscr{M}/V$ is zero, i.e.

$$\mathscr{M} = -V\mathscr{H}/4\pi, \tag{83.35}$$

since a superconductor is perfectly diamagnetic, no magnetic field being able to enter it. (Surface layers are disregarded.) On the other hand, when $\mathscr{H} > \mathscr{H}^*$, $\mathscr{B} = \mathscr{H}$ and so $\mathscr{M} = 0$, disregarding the small normal susceptibility of the metallic crystal.

The transition $2 \to 1$ is reversible and a latent heat λ is associated with it, so that it is a first-order transition. Accordingly, one has at once an equation exactly of the Clapeyron–Clausius type (83.29), i.e. if T^* is the transition temperature,

$$d\mathscr{H}^*/dT^* = -\Delta S/\Delta\mathscr{M}.$$

Since $\Delta S = \lambda/T^*$ and $\Delta\mathscr{M} = V\mathscr{H}^*/4\pi$, the last equation therefore becomes [1*2]

$$d\mathscr{H}^*/dT^* = -4\pi\lambda/VT^*\mathscr{H}^*. \tag{83.36}$$

By differentiating the equation

$$4\pi\Delta S = -V\mathscr{H}^*d\mathscr{H}^*/dT^*,$$

and then setting $\mathscr{H}^* = 0$ one gets for the difference between the specific heats at zero field

$$\Delta C = -\frac{VT^*}{4\pi}\left(\frac{d\mathscr{H}^*}{dT^*}\right)^2, \tag{83.37}$$

the derivative on the right being calculated at $\mathscr{H}^* = 0$. From (83.36) it follows that λ goes to zero with $\mathscr{H}^*$. On the other hand, $d\mathscr{H}^*/dT^*$ is known to remain finite. Evidently one therefore has *for zero field* a state of affairs at the transition temperature T_0^* typical of a second-order transition. The remark following (83.33) is, however, relevant here, since the transition is of the second order only at the isolated point $T = T_0^*$, $\mathscr{H} = 0$, whilst it is of the first order elsewhere.

When the conductor has some shape other than that assumed above, the details of the theory become rather more involved, because one then has to distinguish between the external field $\mathscr{H}_e$ and the internal field $\mathscr{H}_i$.

(*c*) Liquid helium, He^4, provides another example of a transition which is not of the first order. Below about 2·2 °K, at a temperature depending on the pressure, the 'ordinary' helium, He I, will go over into another modification, He II. There is no latent heat involved in this transition, nor is there any change in density. Thus, when He I is cooled it will change abruptly into He II when the transition temperature is reached, without a heterogeneous mixture ever being formed. In this sense, therefore, the two 'phases' cannot coexist—a situation very different from that which obtains in the case of first-order transitions. The inverted commas are intended to emphasize the comparative lack of distinction of the phases implicit in g, s, v, u having the same values for the one as for the other at the transition point. Yet they are physically quite distinct, for quite apart from the measured values of c_P, α and κ being quite different in the two phases, the viscosity and thermal conductivity of He II differ greatly from those of He I.

It would appear, then, that the transition of He^4 just described, the so-called λ-transition, is a second-order transition. This is indeed so. On the other hand, experimental results suggest that c_P, α, and κ all tend to infinity as the transition point is approached from either direction, so that (83.32) then becomes nugatory. They must therefore be modified in circumstances such as these. The salient point is that the mere classification of transitions according to 'order' in Ehrenfest's sense provides an incomplete characterization of the possible types of phase transitions: if a transition is of order m one must also say something about the actual nature of the discontinuity of the mth derivative of g at the transition point. Experimentally it is difficult, if not impossible, to decide whether a plot of c_P against T in which c_P appears to go to infinity at the transition temperature really represents an infinite discontinuity, or merely one which is finite but very large. This kind of difficulty bedevils the empirical aspects of higher-order transitions.

(vii) *Representation of discontinuities.* It is sometimes desirable to write non-differentiable or discontinuous functions in such a way that they may formally be treated as if they were indefinitely differentiable. In suitable circumstances this may be done by introducing the step function $\theta(t)$ and the delta function $\delta(t)$, defined in the usual way:

$$\theta(t) = \begin{cases} 1 & (t > 0) \\ \frac{1}{2} & (t = 0) \\ 0 & (t < 0), \end{cases} \qquad (83.38)$$

$$\delta(t) = 0 \quad (t \neq 0), \qquad \int f(t)\delta(t)\,dt = f(0), \qquad (83.39)$$

where the range of integration includes the origin $t = 0$, and $f(t)$ is any function of t continuous at $t = 0$. The derivative of $\theta(t)$ is just $\delta(t)$. Consider, for example, a first-order transition. The Gibbs function of one mole of a substance is now

$$g = g^1\theta(t) + g^2\theta(-t), \qquad (83.40)$$

where

$$t = T - T^*(P), \qquad (83.41)$$

$T^*(P)$ being the temperature at which, as T increases through it, the phase transition $2 \rightarrow 1$ takes place. Then

$$s = -g_{,T} = s^1\theta(t) + s^2\theta(-t) + (\Delta g)\delta(t).$$

However, only at $t = 0$ does $\delta(t)$ not vanish, but Δg does so just there. The last term on the right therefore vanishes. (If any doubts remain on this score, integrate $\Delta g\delta(t)$ over a small interval containing the origin.) Thus s changes by Δs at $T = T^*$, as required. If one differentiates a second time, one again gets a delta function, with a factor $\Delta s(=\lambda/T^*)$ this time. Thus

$$c_P = c_p^1\theta(t) + c_p^2\theta(-t) + \lambda T^{-1}\delta(t), \qquad (83.42)$$

where the last term on the right neatly exhibits the infinite discontinuity of c_P. Finally, from

$$s = s^1\theta(t) + s^2\theta(-t), \quad v = v^1\theta(t) + v^2\theta(-t)$$

one gets the Clapeyron–Clausius equation at once by means of the usual identity $s_{,P} = -v_{,T}$. With this formalism one may then always write, for instance, $s = \int \tilde{c}_P(P, T)\,dT$ (P = const.) without having to worry about the discontinuities of c_P.

(viii) *Examples of systems with several deformation coordinates.*

(*a*) Most of the theory developed hitherto has been for general systems, that is to say, the number of deformation coordinates has by no means been restricted to be unity. Nevertheless, in applications of the theory the case $n = 2$ has been prevalent. In this subsection two tangible examples of systems with $n > 2$ will therefore be considered very briefly. The first of these concerns magnetic substances, say a paramagnetic gas, or a superconducting (homogeneous, isotropic) medium. $\mathscr{M}$ is the magnetic moment of the medium in the direction of the external field $\mathscr{H}$. To keep the discussion as simple as possible, a situation has here been presupposed such that in the region of volume V occupied by the magnetic substance K, the external field and the magnetization of K are both homogeneous. One may think of K being situated within a uniformly wound solenoid through which a current is passing. Then in order to increase $\mathscr{M}$ by $d\mathscr{M}$ the generator supplying the current has to do an amount of work

$$-dW = d(\mathscr{H}^2 V/8\pi) + \mathscr{H} d\mathscr{M}. \tag{83.43}$$

However, the first term on the right is a total differential, so that when U is defined in the usual manner through the work done in adiabatic transitions, one will have

$$d(U - \mathscr{H}^2 V/8\pi) - \mathscr{H} d\mathscr{M} = 0.$$

Evidently the purely mechanical term $-\mathscr{H}^2 V/8\pi$ will be carried through all stages of calculation. Consequently one may as well understand by U not the energy function as usually defined, but as that energy function less the magnetostatic field energy within the empty solenoid. With this understanding one has eventually, in the usual way,

$$T dS = dU + P dV - \mathscr{H} d\mathscr{M}. \tag{83.44}$$

(If changes in composition were to be admitted, then a term $-\Sigma \mu_i dn_i$ would have to be added on the right.) Here, then, one has a system with two deformation coordinates, viz. $x_1 = V$, $x_2 = -\mathscr{M}$. The Gibbs function for this is therefore

$$G = U - TS + PV - \mathscr{H}\mathscr{M}, \tag{83.45}$$

so that

$$dG(P, \mathscr{H}, T) = -S dT + V dP - \mathscr{M} d\mathscr{H}. \tag{83.46}$$

Equation (83.34) is the same as (83.45) except in so far as the term PV was omitted from the first, on the grounds that its effects were negligible under the circumstances envisaged earlier. In writing down the last few equations the possibility of quasi-static transitions has of course been taken for granted. This means incidentally that there must be no magnetic hysteresis effects.

The integrability conditions on dU, dF, dG, dH give not only the four Maxwell relations but eight further equations of a similar kind, which involve the magnetic quantities. There is little point in reproducing all these here, and one example will suffice. From

$$dH = d(U+PV-\mathscr{H}\mathscr{M}) = TdS+VdP-\mathscr{M}d\mathscr{H}$$

there comes
$$T_{,\mathscr{H}} = -\mathscr{M}_{,S}, \tag{83.47}$$

the coordinates being S, P, $\mathscr{H}$.

Equation (83.47) may be used in investigating the effect on the temperature of K produced by an adiabatic, isobaric change of $\mathscr{H}$, which is also quasi-static. Thus

$$dT = T_{,\mathscr{H}}d\mathscr{H} = -\mathscr{M}_{,S}d\mathscr{H}. \tag{83.48}$$

But $\mathscr{M}_{,S} = \mathscr{M}_{,T}/S_{,T} = T\mathscr{M}_{,T}/C$, where C is the specific heat at constant P and $\mathscr{H}$, so that [1₂]

$$dT = -C^{-1}T\mathscr{M}_{,T}d\mathscr{H}. \tag{83.49}$$

For the sake of illustration, suppose the following conditions to be satisfied: (i) the susceptibility of the body is so small that the distinction between internal and external field may be ignored; (ii) the specific heat C is constant; (iii) the system obeys *Curie's Law*, i.e. $\mathscr{M}$ is a linear function of $\mathscr{H}/T$. Thus

$$\mathscr{M} = \Gamma\mathscr{H}/T, \tag{83.50}$$

where Γ is a constant. Equation (83.49) becomes

$$TdT = \Gamma C^{-1}\mathscr{H}\,d\mathscr{H},$$

which may be integrated to yield

$$T^2-T_0^2 = \Gamma C^{-1}\mathscr{H}^2, \tag{83.51}$$

where T_0 corresponds to zero field. $\delta T = T-T_0$ will be small, so that [1₂*]

$$\delta T = \Gamma\mathscr{H}^2/2CT. \tag{83.52}$$

Since $\Gamma/C > 0$ the temperature may therefore be lowered by reducing the field adiabatically to zero. Although (83.52) is valid at moderate temperatures, it ceases to hold at temperatures of a few degrees K, as must be the case since (83.52) contravenes the Third Law (see also Section 84*c*).

Reverting for a moment to the discussion of superconductivity, (83.36) was classified as being of type 1_2^* rather than of type 1_2 on account of the omission of the term PV from G. However, from (83.46) one gets, on differentiating along the equilibrium surface [*surface*, because there are three coordinates now],

$$-\Delta S\,dT^*+\Delta V\,dP^*-\Delta\mathcal{M}\,d\mathcal{H}^* = 0, \tag{83.53}$$

the asterisk on P^* emphasizing that one is concerned with points on the transition surface. This equation evidently gives rise to three equations of Clapeyron–Clausius type, one of which is (83.36), if the derivative $d\mathcal{H}^*/dT^*$ be understood to refer to constant pressure. Of the other two equations one is the Clapeyron–Clausius equation itself, whilst the other reads

$$\frac{d\mathcal{H}^*}{dP^*} = \frac{4\pi\Delta V}{V^*\mathcal{H}^*}, \tag{83.54}$$

which shows how under isothermal conditions the critical field strength $\mathcal{H}$ depends upon the pressure.

(*b*) The second example of a system with several deformation coordinates is an elastic solid K. When such a solid is deformed by the application of external forces, then, upon these being removed, it will return to its initial configuration. (Otherwise the body would be called plastic.) In practice elastic solids are realizable provided the extent of all deformations is kept below a certain limit. K can thus undergo quasi-static changes and may be regarded as a thermodynamic system.

As usual, the purely mechanical part of the theory will be largely presupposed. However, to clarify the notation a few points will be covered explicitly. A point whose coordinates in the unstrained body are x_i ($i = 1, 2, 3$) shall have the coordinates x_i+u_i in the strained body, so that the u_i are the components of the displacement vector. A point x_i+dx_i neighbouring to the point x_i has the coordinates

$$x_i+dx_i+u_i+du_i = x_i+u_i+dx_i+u_{i,k}dx_k,$$

where summation from 1 to 3 over any repeated index is understood. The final distance dl' between the two points is therefore related to their initial distance dl through

$$dl'^2 = dl^2 + 2u_{i,k}dx_i dx_k + u_{i,k}u_{i,l}dx_k dx_l.$$

If one restricts oneself to small deformations, as will be done here, the last term on the right may be neglected, and then

$$dl'^2 = dl^2 + 2s_{ik}dx_i dx_k, \tag{83.55}$$

where

$$s_{ik} = \tfrac{1}{2}(u_{i,k} + u_{k,i}) \tag{83.56}$$

is the *strain tensor*. s_{ik} has been written in this way to ensure its symmetry. Note that only the symmetric part of $u_{i,k}$ enters into (83.55); the skew-symmetric part of $u_{i,k}$ relates to rigid rotations of the body as a whole, which are of no interest in the context of deformations.

A volume element dV of extension dx_1, dx_2, dx_3, subtended by the three displacements $\delta_{ik}dx_k$ ($i = 1, 2, 3$; $\delta_{ik} = 1$ ($i = k$), $\delta_{ik} = 0$ ($i \neq k$)) becomes the volume element dV' subtended by the three displacements $(\delta_{ik} + u_{i,k})dx_k$, its extension being the determinant of these nine components, i.e. to the required order,

$$dV' = dV(1 + u_{ii}) = dV(1 + s), \tag{83.57}$$

where s is the trace of the strain tensor. The divergence-free part of s_{ik} therefore represents a pure shear.

Let $f_i dV$ be the ith component of all the forces acting on the element dV, so that, to the required order, $\int f_i dV$ is the component of the force acting on any finite part of the body K. Since this part is held in equilibrium by the forces acting over its surface, it must be possible to transform the integral into a surface integral. Consequently there must exist a tensor t_{ik} such that

$$f_i = t_{ik,k}. \tag{83.58}$$

By considering the turning moment of the forces acting on any part of K one can show that t_{ik} is symmetric:

$$t_{ik} = t_{ki}. \tag{83.59}$$

t_{ik} is called the *stress tensor* of K. In the case of a uniform compression from all sides, i.e. a hydrostatic compression,

$$t_{ik} = -P\delta_{ik}. \tag{83.60}$$

Unless a deformation is uniform, i.e. unless the s_{ik} are constant throughout K, the energy density, entropy density and so on will vary from point to point. In that case one may, in analogy with the situation of Section 73, regard every volume element as a metaphase of K. Every metaphase of the kind considered earlier was described by two coordinates, and in *this* sense the system as a whole could be thought of as characterized by two coordinates rather than by an infinite number. In the same way K has here *six* deformation coordinates, since every metaphase is characterized by, say, the six independent components of the strain-tensor and by its temperature. The work done on a volume element δV in a deformation in which s_{ik} changes by ds_{ik} is

$$\delta dW = -t_{ik} ds_{ik} \delta V.$$

Hence if the usual symbols U, S, ... are used to denote densities (e.g. U is the energy per unit volume) one has, according to the usual arguments,

$$T dS = dU - t_{ik} ds_{ik}. \tag{83.61}$$

Then the Gibbs function, for instance, is

$$G(t_{ik}, T) = U - TS - t_{ik} s_{ik}. \tag{83.62}$$

In the case of hydrostatic compression the term $t_{ik}s_{ik}$ becomes, because of (83.60), $-Ps$. According to (83.57) s is the relative increase of volume of the element whose initial volume was V_0. Accordingly, G as given by (83.62) differs from that as previously defined by an additive term $-PV_0$, but this is clearly of no consequence.

From (83.62)

$$dG = -S dT - s_{ik} dt_{ik}, \tag{83.63}$$

and the integrability conditions on this are [1[2]]

$$\partial S/\partial t_{ik} = \partial s_{ik}/\partial T, \quad \partial s_{ik}/\partial t_{lm} = \partial s_{lm}/\partial t_{ik}. \tag{83.64}$$

In carrying out differentiations of this kind the symmetry of t_{ik} and s_{ik} is to be ignored throughout. The first six of the relations (83.64) are generalizations of the third Maxwell relation

$$S_{,P} = -V_{,T}.$$

One may now proceed as before to define all sorts of phenomenological coefficients, and to derive identities between them.

For example, corresponding to C_V one has the specific heat at constant strain $C_{(s)}$, reckoned here per unit volume:

$$C_{(s)} = TS_{,T}\,(s_{ik}\text{ const.}); \tag{83.65}$$

whilst in place of C_P one has the specific heat $C_{(t)}$ at constant stress

$$C_{(t)} = TS_{,T}\,(t_{ik}\text{ const.}). \tag{83.66}$$

Between these there exist relations analogous to (81.10). However, it is hardly necessary to spell such results out in detail. At any rate, the development so far is of general validity, subject to the restriction to small deformations. Even changes of composition can be accommodated in the usual way.

To go further, one may make specific assumptions about the way in which the stresses depend upon the strains. Thus, for small deformations, one may take Hooke's Law to be valid. This states that the stresses set up in K by an isothermal deformation are linear functions of the strains:

$$t_{ik} = a_{iklm}s_{lm}, \tag{83.67}$$

where, at any point, the elastic coefficients a_{iklm} can depend at most upon T. They are symmetric in i, k and in l, m; and on account of the second set of identities (83.64) they are also invariant under the interchange of the first with the second pair of indices. Thus altogether

$$a_{iklm} = a_{kilm} = a_{ikml} = a_{lmik}. \tag{83.68}$$

One is left with 21 independent coefficients, though this number is further reduced by any symmetries the solid may possess. A crystal with cubic symmetry, for example, has only three independent coefficients.

Equation (83.67) is a set of linear equations which may be inverted:

$$s_{ik} = b_{iklm}t_{lm}. \tag{83.69}$$

Then from (83.63, 67)

$$\partial G/\partial t_{ik} = -s_{ik} = -b_{iklm}t_{lm},$$

which may be integrated to give

$$G = G_0 - \tfrac{1}{2}b_{iklm}t_{ik}t_{lm}, \tag{83.70}$$

where G_0, the Gibbs function of the undeformed body, is a function of T only.

For a change which is not isothermal one may have deformations without any stresses appearing (thermal expansion). In that case there must appear on the right-hand side of (83.70) an additional term *linear* in t_{ik}:

$$G = G_0 - \alpha_{ik} t_{ik} - b_{iklm} t_{ik} t_{lm}, \tag{83.71}$$

where $\alpha_{ik}(T)$ is a (symmetric) tensor, since G must be a scalar. Then

$$s_{ik} = \alpha_{ik} + b_{iklm} t_{lm}, \tag{83.72}$$

which replaces (83.69). In free thermal expansion, which is such that there are no stresses at any time,

$$s_{ik} = \alpha_{ik}, \tag{83.73}$$

so that $\alpha_{ik,T}$ is in effect the tensor of linear thermal expansion.

84. The Third Law invoked

(*a*) The Third Law (60.4) requires that

$$\lim_{T\to 0} TS_{,T} = 0, \qquad \lim_{T\to 0} S_{,x_i} = 0 \quad (i = 1, 2, \ldots, n-1). \tag{84.1}$$

The fact that the ideal gas violates the condition $S_{,V} \to 0$ as $T \to 0$ provided just the motivation for introducing the notion of a super-ideal gas in Section 71, and this application of the Third Law need not be dealt with again. Amongst the integrability conditions on dF, as given by (62.5), are the following:

$$S_{,x_i} = P_{i,T}, \tag{84.2}$$

and this set is a generalization of the second of Maxwell's relations (81.5). Hence one has the conditions [13]

$$\lim_{T\to 0} P_{i,T} = 0. \tag{84.3}$$

One may equally well take $P_1, \ldots, P_{n-1}, T$ as coordinates in place of those which occur in (84.1), for the Third Law has been so expressed that no particular choice of coordinates is prescribed. (The singling out of deformation coordinates in the definition of states of least energy is clearly irrelevant here.) Then

$$\lim_{T\to 0} TS_{,T} = 0, \qquad \lim_{T\to 0} S_{,P_i} = 0 \quad (i = 1, 2, \ldots, n-1). \tag{84.4}$$

Using $S_{,P_i} = -x_{i,T}$ it follows that [1₃]

$$\lim_{T\to 0} x_{i,T} = 0. \tag{84.5}$$

These results are relevant to the elastic coefficients of Section 83 (viii):

$$\lim_{T\to 0} a_{iklm,T} = \lim_{T\to 0} \alpha_{ik,T} = 0. \tag{84.6}$$

(*b*) The first members of (84.1) and (84.4) respectively express the vanishing of the specific heat at constant deformation coordinates and constant forces when $T = 0$. It suffices to take $n = 2$. Then [1₃]

$$\lim_{T\to 0} C_V = 0, \qquad \lim_{T\to 0} C_P = 0. \tag{84.7}$$

The individual specific heats go to zero with T more slowly than the difference between them, for each of the three factors on the right of (81.10) goes separately to zero with T. For example, reverting to the super-ideal gas, that is to say (71.15) with $a \neq 0$, it turns out that, whereas C_V and C_P vanish as T^β, $C_P - C_V$ vanishes as $T^{2\beta+1}$ as $T \to 0$. In particular, for Fermi–Dirac gases ($\beta = 1$) [1₁*]

$$C_P \sim \text{const. } T, \quad C_V \sim \text{const. } T, \quad C_P - C_V \sim \text{const. } T^3. \tag{84.8}$$

In general the vanishing of C_V ($T = 0$) is of course not sufficient to ensure that the Third Law is not violated. On the other hand, it may be sufficient when some specific information about the behaviour of the system is at hand. This is so, for instance, in the case of a substance obeying Grüneisen's Law (Section 83 (iv)), for (83.20) shows at once that when C_V vanishes then so does $S_{,V}$.

(*c*) With regard to paramagnetic substances (60.1) requires that

$$\lim_{T\to 0} \mathcal{M}_{,T} = 0. \tag{84.9}$$

Now one sometimes defines an 'ideal paramagnetic' to be a substance such that $\mathcal{M}$ depends on $\mathcal{H}$ and T only in the combination $\mathcal{H}/T$ ($= z$, say), though alternative definitions have been proposed. This is suggested by Curie's Law (Section 83 (viii, *a*)), and by some simple statistical considerations. The existence of such an ideal paramagnetic would, however, contravene the Third Law. Thus, write

$$\mathcal{M} = B(z), \tag{84.10}$$

so that

$$\mathcal{M}_{,T} = -\mathcal{H}^{-1} z B'(z).$$

It will suffice here to suppose that as $T \to 0$ $B(z) \sim az^{\alpha}$ (a, α = const.). Then the vanishing of $zB'(z)$ as $z \to \infty$ requires that $\alpha < 0$. However, $\mathscr{M}$ then tends to infinity as $\mathscr{H}$ is isothermally decreased to zero, which is inadmissible. At sufficiently low temperatures (and in practice these may have to be very low indeed) a phase transition must therefore occur, so that, below the transition temperature, (84.10) is no longer valid.

In conclusion it should be mentioned that an important application of the Third Law occurs in the context of chemical equilibria. This is, however, dealt with more easily later on, after some preliminary work has been carried through. The interested reader will find the details in Section 94.

CHAPTER 11

APPLICATIONS (II)

85. The critical region

(*a*) This final chapter concerns itself with applications of type 2, which, it will be recalled, are just those which require an appeal to the Entropy Principle. In other words, these applications involve, in essence, the contemplation of unnatural states, as set out in detail in Section 69. A suitable first topic is the relevance of the stability condition (70.4) to a system K consisting of a fixed amount of some simple substance; the discussion of Section 79 being thus supplemented to some extent. Here it will be supposed that at most the gaseous phase and the liquid phase are present. K is to be considered as a whole, the coordinates being V and T, as usual. When K is homogeneous the isothermals certainly satisfy the condition $P_{,V} < 0$. On the other hand, when T is less than a certain temperature T_c there exists a range of values of the volume, $V' \leqslant V \leqslant V''$, within which P is exactly constant. V' and V'' depend on T, and $V'' - V'$ is a decreasing function of T. $V'' = V'$ ($= V_c$, say) when $T = T_c$. The straight part of an isothermal corresponds to the situation in which K is heterogeneous.

The point V_c, T_c of R_2 is called the *critical point* $\mathfrak{S}_c$, V_c, T_c being the critical volume and critical temperature respectively, whilst the corresponding pressure P_c is the critical pressure. At $\mathfrak{S}_c$ one evidently has

$$P_{,V} = 0, \quad P_{,VV} = 0, \tag{85.1}$$

and stability will then be assured if

$$P_{,VVV} < 0. \tag{85.2}$$

(Cf. the end of Section 70.) Whether (85.2) is fulfilled in practice is not entirely settled. It is possible that the derivatives of P of order 1, 2, ..., $2n$ vanish, but that of order $2n+1$ does not; or else all derivatives vanish at $\mathfrak{S}_c$, in which case an expansion of F in ascending powers of dV, assumed to exist in Section 70, in fact does not exist. In this context it should be noted that the behaviour of specific heats, isothermal compressibilities, and so forth, as functions of $T - T_c$, predicted under the assumption that the Gibbs functions of the two phases possess Taylor expansions in the vicinity of the critical point seems to be definitely at variance with experimental results.

Empirical equations of state in which P is given as a continuous and differentiable function of V and T are obtained essentially by a process of curve fitting in the two homogeneous regions. Such an equation will naturally fail to reproduce the true state of affairs in the heterogeneous region, since in the latter the 'pseudo-isothermals', i.e. curves T = const., as given by the equation, will lie partly above and partly below the actual isothermal. States lying on the pseudo-isothermal at points where $P_{,V} < 0$, though not completely stable, are experimentally realizable in careful experiments (one has the phenomena of superheating and super-cooling), whilst those at points where $P_{,V} > 0$ are not realizable at all. Nevertheless, on paper one may imagine the system to be taken from V' to V'' along the actual isothermal and back to V' along the pseudo-isothermal, arguing that only the onset of in-stability prevents one from actually carrying out this procedure. Then for this closed isothermal transition $\oint dU = T\oint dS = 0$ and so the position of the actual isothermal is such that $\oint P dV = 0$ (Maxwell's Law of Equal Areas). However, the validity of this argument is admittedly doubtful.

(*b*) An additional remark concerning the passage along the parts of the pseudo-isothermal at which $P_{,V} < 0$ may not be out of place. Suppose a system K, consisting of n moles of a gas C, is compressed quasi-statically and at constant temperature T'. P' shall be the pressure at which a phase change (let it be liquefaction for the sake of argument) normally sets in. The molar Gibbs functions of the gaseous phase and of the liquid phase, μ^1 and μ^2 respectively, depend on P only, T being fixed. When $P < P'$, $\mu^1 < \mu^2$ but μ^1 increases more rapidly with P than μ^2, and at P' the values of the two functions become equal. μ^1 is a regular function of P in a certain finite neighbourhood D of the point $P = P'$, so that $G_{,PP}$ exists, whilst it is supposed here that this derivative, and therefore $P_{,V}$ is negative in D. Homogeneous states corresponding to points of D would, in general, therefore be realizable, were it not for the fact that

$$(n-\xi)\mu^1(P, T') + \xi\mu^2(P, T') < n\mu^1(P, T')$$

when $P > P'$, so that once heterogeneous states are admitted the minimum of G corresponds to $\xi > 0$. (Recall the remark following the derivation of the condition (70.4).) The actual realization of homogeneous states ($P > P'$) evidently requires that the formation of the liquid phase be inhibited for some reason. Such an inhibition may for instance arise in practice from surface tension effects, which have here been left out of account. These will oppose the formation of very small drops of the liquid; but once a drop of sufficient size is formed it must continue to grow until $\xi = n$. The onset of liquefaction will therefore come about only when some fluctuation nullifies the effect of surface tension, the fluctuation being either spontaneous or the result of an external disturbance.

(c) If an empirical equation of state contains just two adjustable constants the value of $Z = PV/RT$ at $\mathfrak{S}_c$ is uniquely determined by (85.1, 2). Van der Waals's equation (79.3) is of this kind. In this case

$$a = 9RT_cV_c/8, \quad b = V_c/3, \quad Z_c = \tfrac{3}{8}. \tag{85.3}$$

The gases O_2, A, CO, N_2 in fact have $Z_c \approx 0.29$, a value which is fairly representative for common gases. The value given by van der Waals's equation is therefore too large. In van der Waals's equation a and b are supposed to be constants. The results just derived remain valid, however, if in the equation a and b are any functions of T, provided one understands a and b in (85.3) to stand for a_c, b_c respectively. This remark is relevant, for instance, to Berthelot's equation, which also has the form (83.29), but with $a = a'/T$ ($a' =$ const.).

If one writes $P = P_cP^*$, $V = V_cV^*$, $T = T_cT^*$, the starred quantities are called reduced pressure, volume, and temperature respectively. The van der Waals equation then reads

$$(P^*+3/V^{*2})(3V^*-1) = 8T^*. \tag{85.4}$$

Two van der Waals gases at the same reduced volume and temperature therefore have the same reduced pressure (Law of Corresponding States).

When one considers that van der Waals's equation was not in the first place intended to do more than to describe small deviations from ideal behaviour, the comparative closeness of the value $\tfrac{3}{8}$ of Z_c to that actually observed is somewhat surprising. Still, one gains the impression that Z_c is rather insensitive to the specific form of a two-parameter equation of state. To illustrate this point consider the virial expansion up to and including the third virial coefficient:

$$Z = 1+B(T)V^{-1}+C(T)V^{-2}. \tag{85.5}$$

In the nature of things one intends this to apply only to situations in which $CV^{-2} \ll BV^{-1} \ll 1$. However, let this reservation be ignored, and pretend that (85.5) is exact. (The actual form of $B(T)$ and $C(T)$ is irrelevant.) Then from (85.1, 5),

$$B_c = -V_c, \quad C_c = \tfrac{1}{3}V_c^2, \quad Z_c = \tfrac{1}{3}, \tag{85.6}$$

so that the value of Z_c is, if anything, better than that provided by van der Waals's equation.

In accordance with the remark following (85.2), suppose now that the derivatives of P of order 1, 2, ..., $2n$ vanish, whilst that of order $2n+1$ does not. To adopt a truncated virial expansion, as was done above, one must include at least all virial coefficients up to the $(2n+1)$th. This ought to be increasingly advantageous with larger and larger n. Let the virial series therefore be truncated after the $(2n+1)$th term:

$$Z = 1+\sum_{r=1}^{2n} Z^{(r+1)}(T)V^{-r}. \tag{85.7}$$

Setting $P = P_c$, $T = T_c$, this may be written

$$V^{2n+1}-(V_c/Z_c)\left[V^{2n}+\sum_{r=1}^{2n} Z_c^{(r+1)}V^{2n-r}\right] = 0.$$

This equation must have the form $(V-V_c)^{2n+1} = 0$, since, by hypothesis, it must have $2n+1$ equal roots. By inspection of the coefficient of V^{2n} it follows that

$$Z_c = 1/(2n+1). \tag{85.8}$$

One arrives at the conclusion that with increasing n the value of Z_c will be more and more at variance with observed values.

(*d*) There are various difficulties which arise in practice in making measurements in the region of the critical point. One of these comes from the vanishing of $P_{,V}$ at $\mathfrak{S}_c$. Suppose the gas is contained in a tube of length h, situated vertically in the gravitational field. Equation (73.6) then reads

$$VP_{,V}dV = -Mg\,dz, \tag{85.9}$$

where $V = V(z)$ is the molar volume. To be quite specific, suppose that $T = T_c$ and that things are so arranged that at a height z_0 above the lower end of the tube the density of the gas has its critical value. Further, let the equation of state (85.5) be adopted. Then (85.9) may be integrated:

$$\ln V - 2B_c V^{-1} - \tfrac{3}{2}C_c V^{-2} = \alpha z + \text{const.}, \qquad (85.10)$$

where $\alpha = Mg/RT$. The constant of integration is to be chosen so that $V = V_c$ when $z = z_0$. Write $V = V_c(1+\eta)$, $\eta \ll 1$, and use (85.6). Then (85.10) becomes, on neglecting powers of η exceeding the third,

$$\eta^3 = 3\alpha(z-z_0). \qquad (85.11)$$

By way of example, take the case of oxygen O_2, for which $M = 32$, $T_c = 154°K$. Then $\eta \approx 0{\cdot}02(z-z_0)^{\frac{1}{3}}$. There is therefore a fairly rapid change in density near $z = z_0$ on account of the presence of the gravitational field. In fact, the density will change by 1 % in a layer whose thickness is of the order of 1 mm. At the same temperature, if the behaviour of O_2 were that of an ideal gas, the same change of density would occur only in a layer several thousand centimetres thick. Finally, though the last member of (76.13) is not valid at the critical point, it shows that large density fluctuations will occur at $\mathfrak{S}_c$, and these should reveal themselves optically. This is in fact the case: one has the phenomenon of 'critical opalescence'.

86. Heterogeneous equilibrium of systems of one constituent

A system K shall consist of n^k moles of a single substance C in the kth phase, $k = 1, 2, \ldots, p$. Certain aspects of such a system were already considered in Section 83(v). The results obtained there were regarded as of type 1, being based essentially on the exposition of Section 68. In particular the molar Gibbs functions of C in the various phases must be equal:

$$g^1 = g^2 = \ldots = g^p, \qquad (86.1)$$

and further

$$p \leqslant 3. \qquad (86.2)$$

The investigation of the equilibrium of K should, however, include an examination of its stability, that is to say, it should be undertaken from the point of view of Section 69. This may be done by comparing the actual state of K with neighbouring unnatural states, differing from the former in respect of the mole numbers and the values of external coordinates of the various phases. K is taken to be isolated, so that (69.2) is relevant, and the total volume V, energy U, and quantity n of C remain fixed. Thus

$$\Sigma n^k u^k = U, \quad \Sigma n^k v^k = V, \quad \Sigma n^k = n. \tag{86.3}$$

It is instructive not to assume from the outset that the pressures P^k and temperatures T^l of the various phases are equal. It is convenient to write $du^k = \xi_k$, $dv^k = \eta_k$, $dn^k = \zeta_k$ for the variations of the coordinates u^k, v^k, n^k. Then, correct to the second order, the variation δS of the entropy of K is

$$\delta S = \Sigma[n^k(ds^k + d^2s^k) + \zeta_k s^k + \zeta_k ds^k]. \tag{86.4}$$

The variations of U, V, and n vanish, i.e.

$$\delta U = \Sigma(n^k\xi_k + u^k\zeta_k + \xi_k\zeta_k) = 0, \tag{86.5}$$

$$\delta V = \Sigma(n^k\eta_k + v^k\zeta_k + \eta_k\zeta_k) = 0, \tag{86.6}$$

$$\delta n = \Sigma\zeta_k = 0. \tag{86.7}$$

The first-order terms of δS must vanish, subject to the constraints (86.5–7), the second-order terms being omitted from these for the time being. Using the method of Lagrangian multipliers, the condition $\delta S + \lambda\delta U + \mu\delta V + \nu\delta n = 0$ gives

$$1/T^k + \lambda = 0, \quad P^k/T^k + \mu = 0,$$

$$s^k + \lambda u^k + \mu v^k + \nu = 0. \tag{86.8}$$

Accordingly, all temperatures have a common value, T, say $(= -1/\lambda)$, whilst all pressures have a common value, P, say $(= \mu/\lambda)$. The third member of (86.8) then shows that all molar Gibbs functions have a common value, g say $(= -\nu/\lambda)$.

So far one has merely recovered known results. Proceeding now to the investigation of the second-order terms, add

$$\lambda\delta U + \mu\delta V + \nu\delta n = (-\delta U - P\delta V + g\delta n)/T$$

to the right-hand member of (86.4), δU, δV, δn being given by (86.5–7). The first-order terms of course disappear from δS, and the only remaining second-order term is

$$\delta S = \Sigma n^k d^2 s^k. \tag{86.9}$$

Consider therefore the second differential d^2s, suppressing superscripts for the moment:

$$d^2s = \tfrac{1}{2}(s_{,uu}\xi^2 + 2s_{,uv}\xi\eta + s_{,vv}\eta^2)$$
$$= \tfrac{1}{2}[\xi d(s_{,u}) + \eta d(s_{,v})].$$

Now $s_{,u} = 1/T$ and $s_{,v} = P/T$, whilst, if one now takes T as independent variable in place of u, $\xi = c_V dT + u_{,v} dv$, then

$$2d^2s = -(c_V dT + u_{,v} dv) dT/T^2 + [(P/T)_{,T} dT + (P/T)_{,v} dv] dv.$$

The second term on the right cancels the third, in view of (81.3), so that finally

$$2d^2s = -c_V(dT/T)^2 + T^{-1}P_{,v}(dv)^2 < 0, \tag{86.10}$$

since $c_V > 0$ and $P_{,v} < 0$. Accordingly every individual term of (86.9) is negative, i.e.

$$d^2S < 0, \tag{86.11}$$

so that the extremum of S is indeed a maximum.

The analysis just undertaken is a good example of the straightforward approach which bases itself upon (69.2). However, a full treatment of the problem in which due attention is paid to higher order transitions and critical points, is quite complex. The interested reader may refer to the following paper: L. Tisza, *Ann. of Phys.* **13** (1961), 1.

87. The phase rule for inert systems

(*a*) According to (65.3) the total number of coordinates of a general system K with n external coordinates, p phases, and z constituents is $N = n + pz$. In any variation of the coordinates

$$dG = \sum_{i=1}^{n-1} x_i dP_i - S dT + \sum_{k=1}^{p} \sum_{i=1}^{z} \mu_i^k dn_i^k. \tag{87.1}$$

Suppose first that there are no partitions within K, and that the free passage of the various constituents between the several phases is possible. Equation (69.9) requires that

$$\sum_k \sum_i \mu_i^k dn_i^k = 0. \tag{87.2}$$

In the absence of chemical reactions the dn_i^k are subject only to the conditions arising from the closedness of K, i.e.

$$\sum_k dn_i^k = 0 \quad (i = 1, \ldots, z). \tag{87.3}$$

If the ith of (87.3) be supplied with a Lagrangian multiplier λ_i, one sees that now

$$\sum_k \sum_i (\mu_i^k + \lambda_i) dn_i^k = 0 \tag{87.4}$$

for *arbitrary* dn_i^k, i.e. one must have

$$\mu_i^k + \lambda_i = 0 \quad (i = 1, \ldots, z;\ k = 1, \ldots, p). \tag{87.5}$$

Accordingly the chemical potentials of any given constituent C_i have the same values in all phases.

Now the chemical potential μ_i^k of C_i in the kth phase depends only on the concentrations c_j^k of the various constituents in that phase, and there are $z-1$ of these concentrations, or $p(z-1)$ of them for K as a whole. Equations (87.5) are therefore a set of $z(p-1)$ equations for the $p(z-1)$ concentrations and n external coordinates, that is, for $n+p(z-1)$ unknown quantities in all. Accordingly the number f of variables (the 'number of degrees of freedom' of K) to which arbitrary values may be assigned is [2₂]

$$f = z - p + n. \tag{87.6}$$

It must be clearly understood that the actual masses of the phases are here regarded as irrelevant. In other words, one is concerned only with the values of intensive variables. In the most common situation one has merely P and T as external coordinates, and then

$$f = z - p + 2. \tag{87.7}$$

Equation (87.6) constitutes the *Gibbs Phase Rule* for inert systems. It may be noted that if any of the c_i^k are zero *a priori* then a corresponding number of conditions is absent from (87.5), and (87.6) is unaffected.

If K contains internal diathermic partitions, these may be fixed or movable, impermeable or semipermeable. f will then have a greater value than that given by (87.7). For, on the one hand, corresponding to any fixed membrane there will be the absence of the condition that the pressures in the phases separated by it must be equal; whilst, corresponding to any semipermeable membrane, equality of the chemical potentials of C_i on either side of it obtains only if the membrane is permeable to C_i. It may be remarked that when P and T have the same fixed values on both sides of a partition impermeable to C_i, separating phases k and l, and given that

$\mu_i^k > \mu_i^l$, the removal of the membrane will cause the transfer of C_i from the kth to the lth phase. This transfer from the phase in which the chemical potential of C_i is higher to that in which it is lower may be compared with the transfer of energy between systems initially at different temperatures.

The phase rule combines simplicity with generality in a very satisfying way. It is out of the question to discuss it in any further detail here: whole text-books have been devoted to it. One or two simple examples must suffice. One notes at once from (87.7) that, under the circumstances to which it relates, at most $z+2$ phases of a z-constituent system can coexist. This is a generalization of the result of Section 83(v) that no more than three phases of a one-constituent system can coexist. Again, consider a solution of NaCl in water which is being boiled at constant pressure P. Here $z = p = 2$, and, since P is fixed, one has $f = 1$. Thus this single degree of freedom might be taken to be the concentration c of NaCl in the solution, and evidently the boiling-point T is determined by it. As c increases so does T. However, at some stage the concentration becomes so large that the NaCl crystallizes, and the appearance of a third phase reduces f to zero. As boiling continues c, and therefore T, must remain constant, crystallization and boiling off of water proceeding at just such rates as to ensure the constancy of c.

(*b*) It is instructive to reword the argument leading to (87.7). Thus, focusing attention from the outset on the intensive variables μ_i^k, P, T, there are $z+2$ of these, since every constituent has the same chemical potential in every phase. Taking into account the p Gibbs–Duhem identities (81.20), one arrives at once at the previous conclusion that $f = z-p+2$. The inclusion of surface phases cannot affect this result. One can also see this formally in the following way. Because of (74.3) one has in effect the same number of chemical potentials as before. Corresponding to every surface phase Σ^σ one has an additional Gibbs–Duhem identity. At the same time one has, however, an additional intensive variable, namely the surface tension γ^σ which occurs in this identity. The number of degrees of freedom is therefore the same as if all surface phases had been ignored.

88. Phase rule for chemically active systems

Equation (87.6) was derived under the assumption that the system under consideration was inert. This restriction is now to be abandoned, that is to say, any number of chemical reactions are permitted to occur. Recall Section 68(*b*), where R independent chemical reactions in a single phase were contemplated. The reactions between the constituents were represented by (68.11), i.e.

$$\sum_i \nu_i^{(r)} C_i = 0 \quad (r = 1, \ldots, R < z). \tag{88.1}$$

The mutual independence of the reactions means that there is no equation of the set (88.1) which can be derived by linear combination from the remaining equations of the set. This condition is satisfied if and only if not all R-rowed determinants formed from the $z \times R$ matrix of the coefficients $\nu_i^{(r)}$ vanish. Then any change $dn_i^{(r)}$ of the mole numbers, as a consequence of the rth reaction, is characterized by

$$dn_1^{(r)}/\nu_1^{(r)} = dn_2^{(r)}/\nu_2^{(r)} = \ldots = d\xi^{(r)}, \tag{88.2}$$

$\xi^{(r)}$ being the degree of advancement of the rth reaction.

Generalization to several phases is easily achieved. In place of (88.1) one has

$$\sum_k \sum_i \nu_i^{k(r)} C_i = 0 \quad (r = 1, \ldots, R), \tag{88.3}$$

where, as before, some of the stoichiometric coefficients $\nu_i^{k(r)}$ may be zero. Then, in place of (88.2)

$$dn_i^{k(r)} = \nu_i^{k(r)} d\xi^{(r)}, \tag{88.4}$$

and chemical equilibrium is characterized by

$$\sum_k \sum_i \mu_i^k \nu_i^{k(r)} = 0 \quad (r = 1, \ldots, R). \tag{88.5}$$

It will be noticed that formally (88.5) also covers the mere transfer between phases of a constituent, C_i say, which is not 'reacting' (i.e. undergoing chemical change), namely, whenever in such an equation the only non-zero stoichiometric coefficients are those belonging to C_i. However, for the time being such equations are to be understood as excluded, the corresponding physical process having already been taken into account.

The phase rule relevant to the present situation may be arrived at by the same arguments as that of Section 87. The only modification is that there are now the additional R relations (88.5) which have to be satisfied by the μ_i^k. There follows the required result [2₂]

$$f = z-p+n-R, \tag{88.6}$$

or in the most common case $n = 2$,

$$f = z-p+2-R. \tag{88.7}$$

Equation (88.6) is unaffected by conditions of insolubility. If C_i is insoluble in the lth phase, i.e. not contained in it, the lth of the conditions (87.5) will be absent. At the same time the concentration c_i^l is known to be zero. Thus the reduction of the number of unknowns is just counterbalanced by the reduction of the number of equations of condition.

On occasion the number of degrees of freedom is further reduced by the presence of certain additional conditions imposed from the start upon some of the concentrations. This arises, for instance, when the electrolytic dissociation of a constituent in a solution is regarded as a chemical reaction. Thus if C_i dissociates into some ionic constituents C_i^+ and C_i^-, the concentrations of C_i^+ and C_i^- must be equal to one another. In fact one has here a condition of electrical neutrality. At any rate, if there are R' additional conditions, one has in place of (88.6)

$$f = z-p+n-R-R'. \tag{88.8}$$

One sometimes refers to the number

$$z^* = z-R-R', \tag{88.9}$$

as the 'number of independent constituents' of K. Then (88.8) becomes superficially more like (87.6), i.e.

$$f = z^*-p+n. \tag{88.10}$$

89. Further remarks on the stability of general systems

The stability of a general system with respect to transfer of matter between phases may, in principle, be investigated without difficulty. To begin with, it is now convenient to represent the mere transfer of constituents between phases also by equations of the type of (88.4). It will be recalled that the possibility of doing so was already pointed out just after (88.5). Then every process with respect to which stability is in question is characterized by a parameter $\xi^{(r)}$, there being altogether q, say, of these. Then, from (69.9), stability requires that

$$\sum_{s,t=1}^{q} \Gamma_{st}\, d\xi^{(s)}\, d\xi^{(t)} \geqslant 0, \tag{89.1}$$

where $\Gamma_{st} = \partial^2 G/\partial\xi^{(s)}\partial\xi^{(t)}$, this condition being essentially the same as (70.6). The equality sign in (89.1) is to obtain only when $d\xi^{(s)} = 0$ for all s.

Now let Γ_q be the determinant of the matrix of the coefficients of the quadratic form on the left of (89.1). Further, Γ_p shall be the determinant obtained from Γ_q by striking out all but the first p of its rows and columns. Then a necessary and sufficient condition for (89.1) to hold is that

$$\Gamma_p > 0 \quad (p = 1, 2, \ldots, q). \tag{89.2}$$

Having already considered the equilibrium of one component distributed over different phases, a simple but useful illustration in the present context is that of the equilibrium of a single chemically active phase. This shall be a gaseous phase, and conditions will be supposed to be such that stability with respect to the formation of liquid or solid phases may be taken for granted. For a start, let there be a single chemical reaction, ξ being its degree of advancement. Equation (89.2) reduces to

$$\Gamma_1 = G_{,\xi\xi} > 0. \tag{89.3}$$

Now
$$G_{,\xi} = \sum_{i=1}^{z} G_{,n_i} n_{i,\xi} = \sum_i \mu_i \nu_i,$$

whence
$$\Gamma_1 = \sum_{i,j} \mu_{ij} \nu_i \nu_j, \tag{89.4}$$

with
$$\mu_{ij} = \mu_{i,n_j}. \tag{89.5}$$

Note that μ_{ij} is symmetric, in view of (81.17). Stability therefore requires that the expression on the right of (89.4) shall be positive. Recalling (66.9) this may be written as

$$\Gamma_1 = -\frac{1}{2}\sum_{i,j} \mu_{ij} n_i n_j (\nu_i/n_i - \nu_j/n_j)^2. \tag{89.6}$$

Hence a *sufficient* condition for (89.3) to hold is that

$$\mu_{ij} < 0 \quad (i \neq j). \tag{89.7}$$

This condition is satisfied for a mixture of ideal gases, since in that case one has from (82.16)

$$\mu_{ij} = -RT/n \quad (i \neq j). \tag{89.8}$$

When there are several chemical reactions one may proceed in much the same way. Recalling (68.13), one has in the first place

$$\Gamma_{st} = \sum_{i,j} \mu_{ij} \nu_i^{(s)} \nu_j^{(t)}. \tag{89.9}$$

This may be rewritten in a form which generalizes (89.6):

$$\Gamma_{st} = -\frac{1}{2}\sum_{i,j} \mu_{ij} n_i n_j (\nu_i^{(s)}/n_i - \nu_j^{(s)}/n_j)(\nu_i^{(t)}/n_i - \nu_j^{(t)}/n_j), \tag{89.10}$$

and these are then subject to the conditions (89.2). When the gases are ideal (89.8) applies. Then Γ_{st} takes the generic form

$$\Gamma_{st} = \sum_{i,j} \chi_{ij}^{(s)} \chi_{ij}^{(t)},$$

and one can show that (89.2) is satisfied.

If the gases are not ideal (89.8) must be modified. When it suffices to include only the second virial coefficient one gets from (82.28)

$$\mu_{ij} = -RT/n + (2P/n) \sum_{a,b} [B_{ab} - (B_{ai} + B_{aj}) + B_{ij}] c_a c_b \quad (i \neq j). \tag{89.11}$$

Since the additional terms here must be small, their presence will not affect previous conclusions.

90. Law of Mass Action for ideal gases: one reaction

(*a*) Let there be one chemical reaction (65.1) in a mixture of z ideal gases $C_1, \ldots, C_z$. In equilibrium the condition (68.5) must be satisfied:

$$\sum_{i=1}^{z} \mu_i \nu_i = 0. \tag{90.1}$$

The chemical potential μ_i of C_i is given by (82.16). Expressed in terms of the partial pressure P_i it has the generic form

$$\mu_i = \chi_i(T) + RT \ln P_i, \tag{90.2}$$

where, in view of (82.15),

$$\chi_i(T) = \gamma_i(T) - s_{1i} T + u_{1i}. \tag{90.3}$$

That γ_i depends only on T follows from its definition, if (81.9) and (82.2) are taken into account. When (90.2) is inserted in (90.1) there appears the sum $\Sigma \nu_i \chi_i$ which is a function of T alone, and one defines the function $K(T)$ through

$$\sum_{i=1}^{z} \nu_i \chi_i = -RT \ln K(T). \tag{90.4}$$

Then [2$_2^*$]

$$\prod_{i=1}^{z} P_i^{\nu_i} = K(T). \tag{90.5}$$

$K(T)$ is called the *equilibrium constant* (unqualified) for the reaction in question. If one introduces the concentrations in place of the partial pressures, one gets

$$\prod_{i=1}^{z} c_i^{\nu_i} = K_c(P, T). \tag{90.6}$$

The equilibrium constant K_c [constant for fixed P and T, that is to say] depends now also on P. In fact

$$K_c(P, T) = P^{-\nu} K(T), \tag{90.7}$$

with

$$\nu = \sum_{i=1}^{z} \nu_i. \tag{90.8}$$

Equation (90.5) (or (90.6)) expresses the *Law of Mass Action* for ideal gas mixtures in the presence of a single reaction.

Equation (90.6) is of course no more than a rewritten version of (68.7), the generic form of the chemical potentials being known here. To be more explicit one may introduce the degree of advancement ξ into (90.6). Using (68.6) one has

$$c_i = (n_{i0} + \nu_i \xi)/(n_0 + \nu \xi). \tag{90.9}$$

Then

$$(n_0 + \nu\xi)^{-\nu} \prod_i (n_{i0} + \nu_i \xi)^{\nu_i} = P^{-\nu} K(T). \tag{90.10}$$

By way of illustration, consider the dissociation of one mole of a compound of the type A_2 into two like constituents A:

$$A_2 \rightarrow 2A. \tag{90.11}$$

Evidently $\nu_1 = -1$, $\nu_2 = 2$, $\nu = 1$, $n_{10} = 1$, $n_{20} = 0$, $n_0 = 1$. With these values (90.10) becomes

$$4\xi^2/(1-\xi^2) = K(T)/P, \tag{90.12}$$

It follows at once that a decrease of pressure advances the degree of dissociation.

(*b*) The consideration of general heterogeneous reactions is out of place here, since it would lead too far into the details of chemical thermodynamics. There is, however, a special situation which fits into the present context, namely when an ideal gas reaction between constituents $C_1, \ldots, C_2$ takes place in the presence of the solid phases of some of these constituents. If C_j is one of the latter, its chemical potential is virtually independent of P. It follows at once that the equilibrium equation now has the form

$$\prod_i{}' P_i^{\nu_i} = K(T), \tag{90.13}$$

where the prime indicates that the product is taken over only those constituents which are present in the gaseous phase alone. $K(T)$ is given by

$$-RT\ln K(T) = \sum_i{}' \nu_i \chi_i(T) + \sum_i{}'' \nu_i \mu_i(T), \tag{90.14}$$

where Σ' denotes the sum over those gases whose partial pressures appear on the left of (90.13), whilst the sum Σ'' is extended over the solids present.

91. Law of Mass Action for ideal gases: several reactions

The situation considered in the previous section may be generalized by admitting the simultaneous presence of several independent reactions, say R of them in all. Then one has the R conditions (68.16):

$$\sum_{i=1}^{z} \mu_i \nu_i^{(r)} = 0 \quad (r = 1, \ldots, R), \tag{91.1}$$

in place of the single condition (90.1). Proceeding as before, define

$$\sum_i \nu_i^{(r)} \chi_i = -RT\ln K^{(r)}(T). \tag{91.2}$$

Then, in place of (90.5), [2*2]

$$\prod_{i=1}^{z} P_i^{\nu_i(r)} = K^{(r)}(T) \quad (r = 1, \ldots, R), \tag{91.3}$$

and these R equations must hold simultaneously. Recalling (68.13), one then has from (91.3) R equations for the R degrees of advancement $\xi^{(r)}$:

$$(n_0 + \sum_s \nu^{(s)}\xi^{(s)})^{-\nu^{(r)}} \prod_i (n_{i0} + \sum_s \nu_i^{(s)}\xi^{(s)})^{\nu_i(r)} = P^{-\nu^{(r)}} K^{(r)}(T), \tag{91.4}$$

where $\nu^{(r)} = \sum_i \nu_i^{(r)}$. Any reaction *dependent* upon the independent reactions (68.11) has an equilibrium constant which is a function of $K^{(1)}, \ldots, K^{(R)}$ alone. The stoichiometric equation of a dependent reaction has the form

$$\sum_i \nu_i' C_i = 0,$$

where

$$\nu_i' = \sum_r \alpha_r \nu_i^{(r)}, \tag{91.5}$$

the integers $\alpha_1, \ldots, \alpha_R$ being freely disposable. With this reaction the equation

$$\prod_i P_i^{\nu_i'} = K'(T)$$

is associated, and

$$\ln K'(T) = \sum_r \alpha_r \ln K^{(r)}(T). \tag{91.6}$$

92. Ideal systems and their equilibrium constants

(*a*) The Law of Mass Action as expressed by (90.6) has so simple a form essentially because the concentrations c_i entered into the chemical potential only through the additive term $RT\ln c_i$. One may therefore appropriately go beyond the theory relating to a mixture of ideal gases by defining a class of *ideal systems*. These will, in the first instance, be taken to be homogeneous. A (homogeneous) system is ideal if, for every i,

$$\mu_i = \bar{\mu}_i(P, T) + RT\ln c_i. \tag{92.1}$$

A mixture of ideal gases evidently constitutes an ideal system, and in this case $\bar{\mu}_i(P, T)$ is the chemical potential of the pure constituent C_i ($i = 1, \ldots, z$). Any mixture for which $\bar{\mu}_i$ has this meaning is an *ideal mixture*. A different situation obtains in the case of ideal *dilute solutions*. Here all but one of the constituents are present in very small amounts, say $c_i \ll 1$ ($i > 1$), whilst c_1 is nearly unity: C_1 is the 'solvent' and $C_2, \ldots, C_z$ are the 'solutes'. Then $\bar{\mu}_1$ is again the chemical potential of the pure constituent C_1, but, $\bar{\mu}_2, \ldots, \bar{\mu}_z$ depend also on the nature of the solvent. The dilution necessary to ensure ideality of an actual solution depends upon the nature of the constituents. In the case of electrolytes the solution may have to be very dilute indeed.

If one writes

$$\Sigma \nu_i \bar{\mu}_i = -RT \ln K_c(P, T), \tag{92.2}$$

one again gets, in the presence of one chemical reaction, an equation of the simple form (90.6) [2*2]:

$$\prod_{i=1}^{z} c_i^{\nu_i} = K_c(P, T). \tag{92.3}$$

Different ideal systems are distinguished by the form of the functions $\bar{\mu}_i$. In general $\bar{\mu}_i$ need by no means depend logarithmically upon P, in which case K_c will then not be proportional to a power of P.

(*b*) The constant K_c in (92.2) will be referred to as an *ideal equilibrium constant*, to emphasize that it characterizes the equilibrium of an active *ideal* system. Its dependence upon P and T may be investigated by forming its derivatives with respect to these coordinates. In the first case one has to consider $\bar{\mu}_{i,P}$. Because of (92.1) this is the same as

$$\mu_{i,P} = G_{,n_i P} = V_{,n_i} = v_i. \tag{92.4}$$

v_i is the partial molar volume, i.e. (for an ideal system) the increase in volume of the mixture consequent upon the addition of one mole of C_i at constant P and T. Then $\Sigma \nu_i v_i$ ($=\Delta V$, say) is the increase in volume of the mixture when ν_i moles of C_i ($i = 1, \ldots, z$) are transformed in the chemical reaction proceeding in the direction implied by the stoichiometric equation (65.1), the increase being reckoned at given P and T. Thus finally, from (92.2),

$$(\ln K_c)_{,P} = -\Delta V/RT. \tag{92.5}$$

In the case of a mixture of ideal gases $\Sigma \nu_i v_i = \nu RT/P$, so that the right-hand member of (92.5) becomes $-\nu P$, in harmony with (90.6, 7).

As regards the temperature variation of K_c, one requires first the derivative $(\bar{\mu}_i/T)_{,T}$. Because of (92.1) this is

$$(\mu_i/T)_{,T} = (G/T)_{,ni\,T} = -(H/T^2)_{,ni} = -h_i/T^2, \quad (92.6)$$

where h_i is the partial molar enthalpy of C_i. Now the heat of reaction at constant pressure, Λ say, is the heat generated as the result of the chemical transformation, at constant pressure, of ν_i moles of C_i ($i = 1, \ldots, z$), the reaction proceeding, as before, in the direction implied by its stoichiometric equation. Then

$$\Lambda = -Q = -\Sigma\nu_i h_i, \quad (92.7)$$

keeping in mind the constancy of P. One thus arrives at the equation of van't Hoff:

$$(\ln K_c)_{,T} = -\Lambda/RT^2. \quad (92.8)$$

Equations (92.5) and (92.8) thus exhibit the derivatives of $\ln K_c$ in terms of quantities directly accessible to experiment.

(*c*) In the case of a mixture of ideal gases K_c, or, what then comes to the same thing, K may be written down explicitly. Merely for the sake of illustration, suppose that the specific heats c_{Pi} of the constituents are constant in the temperature range T_1, T. Define constants A, Γ, η as follows:

$$\left.\begin{aligned} \Gamma = \Sigma\nu_i c_{Pi}/R, \quad \eta &= \Gamma T_1 - \Sigma\nu_i u_{1i}/R, \\ \ln A &= \Sigma\nu_i s_{1i}/R - \Gamma(1+\ln T_1). \end{aligned}\right\} \quad (92.9)$$

Then from (90.4), using the results of Section 82(ii), [2_2^*]

$$K(T) = AT^{\Gamma}e^{\eta/T}. \quad (92.10)$$

A and η evidently cannot depend on T_1, for this temperature can be chosen arbitrarily (at any rate within a certain range) and $K(T)$ cannot depend on this choice. (The topic of this subsection will be taken up again in Section 94).

(*d*) A heterogeneous system is ideal if each of its phases is ideal, i.e. if

$$\mu_i^k = \bar{\mu}_i^k(P, T) + RT\ln c_i^k \quad (i = 1, \ldots, z;\ k = 1, \ldots, p). \quad (92.11)$$

As in Section 89, transfer of constituents between phases will also be described by standard stoichiometric equations. For instance the transfer of C_i from the kth to the lth phase has the equation

$$-\nu_i^k C_i + \nu_i^l C_i = 0,$$

and there will be a corresponding degree of advancement, of which there shall be R in all. Everything goes through in much the same way as before. Write, by way of generalization of (92.2),

$$\sum_k \sum_i \nu_i^{k(r)} \overline{\mu}_i^k = -RT \ln K_c^{(r)}(P, T). \tag{92.12}$$

Then one has the simultaneous equilibrium conditions [2_2^*]

$$\prod_{k=1}^{p} \prod_{i=1}^{z} (c_i^k)^{\nu_i^{k(r)}} = K_c^{(r)}(P, T) \quad (r = 1, \ldots, R). \tag{92.13}$$

93. The equilibrium condition for real gases

When the imperfections of gas mixtures are taken into account various devices are used to retain a certain formal (though in practice largely chimerical) simplicity. Thus, let attention be focused first on (82.16). When the mixture is not ideal, one may write

$$\mu_i = g_i^*(P, T) + RT \ln a_i, \tag{93.1}$$

where the asterisk has the significance assigned to it in Section 82 (iii), so that g_1^* is the chemical potential of C_i by itself, calculated as if this constituent were ideal. Equation (93.1) is simply the definition of the quantity a_i, called the *activity* of C_i. Evidently as P tends to zero the effects of all terms but the first of the virial expansion become negligible, and so

$$\lim_{P \to 0} a_i / c_i = 1. \tag{93.2}$$

In general a_i is a function of P, T and all the concentrations. The quantity K_c defined by

$$-RT \ln K_c(P, T) = \Sigma \nu_i g_i^* \tag{93.3}$$

is just that which occurs on the right of (90.6): and the generalization of the Law of Mass Action (90.6) appropriate to the present situation is [2_2^*]

$$\prod_{i=1}^{z} a_i^{\nu_i} = K_c(P, T), \tag{93.4}$$

The simplicity exhibited by this equation has been achieved by adopting a prescription which includes the effects of all imperfections in the activities, at the same time retaining unchanged that term of μ_i which does not depend upon the concentrations. In place of (92.8) one has

$$(\ln K_c)_{,T} = -\Lambda^*/RT^2, \tag{93.5}$$

i.e. Λ^* is not the actual heat of reaction (at pressure P) but the heat of reaction measured at a pressure so low that the imperfections of the mixture can be neglected.

Instead of retaining the formal structure of (82.16) one may retain that of (90.2) by defining a quantity ϖ_j, the *fugacity* of C_i, through the relation

$$\mu_i = \chi_i^*(T) + RT\ln\varpi_i, \tag{93.6}$$

where

$$\lim_{P\to 0} (\varpi_i/c_iP) = 1. \tag{93.7}$$

If $K(T)$ is defined as in (90.4), i.e. $\Sigma\nu_i\chi_i^* = -RT\ln K$, then (90.5) generalizes to

$$\prod_{i=1}^{z} \varpi_i^{\nu_i} = K(T). \tag{93.8}$$

This is of course entirely equivalent to (93.4). If one defines *activity coefficients* γ_i through

$$\gamma_i = a_i/c_i \tag{93.9}$$

(cf. (93.2)), one easily convinces oneself by comparing (93.1) with (93.6) that

$$\varpi_i/P_i = \gamma_i, \tag{93.10}$$

where the partial pressure P_i is, by definition, c_iP, as usual. The generalization of (90.5) to

$$\prod_{i=1}^{z} P_i^{\nu_i} = K^\# \tag{93.11}$$

achieves nothing but to shift the complications arising from imperfections from the left-hand member of (93.8) to the right-hand member of (93.11), $K^\#$ being now a function of P, T and the concentrations. In fact

$$K^\# = K(T)\prod_i \gamma_i^{-\nu_i}. \tag{93.12}$$

An explicit example of activity coefficients may be read off from (82.28) which relates to a situation in which the imperfections are sufficiently accounted for by the second virial coefficients. Thus

$$a_i = c_i\{1 + (P/RT)\sum_{j,k}(2B_{ij} - B_{jk})c_jc_k\}, \tag{93.13}$$

since only terms linear in the B_{jk} are to be retained. In the second term within the braces the concentrations c_l may be replaced by their values c_l^* which they would have in the absence of imperfections. With this understanding, write

$$a_i = c_i(1+\alpha_i). \tag{93.14}$$

Further, set

$$c_i = c_i^*(1+\eta_i). \tag{93.15}$$

Taking (90.6) into account, (93.4) then gives

$$\Sigma\nu_i(\eta_i+\alpha_i) = 0. \tag{93.16}$$

In the absence of imperfections the degree of advancement would have the known value ξ^*. Then η_i will depend linearly on $\xi-\xi^*$ since the presence of higher powers of $\xi-\xi^*$ would be equivalent to retaining higher powers of the second virial coefficients. Thus

$$\eta_i = [(n_0\nu_i-\nu n_{i0})/n^*n_i^*]\,(\xi-\xi^*) = \kappa_i(\xi-\xi^*), \tag{93.17}$$

say, and then finally

$$\xi-\xi^* = -\Sigma\nu_i\alpha_i/\Sigma\nu_i\kappa_i, \tag{93.18}$$

all quantities on the right being known explicitly. When the imperfections of the reactants are substantial, the details of the solution of (93.4) become much more unwieldy. However, in principle one has of course always merely a complicated algebraic equation for ξ.

94. Calorimetric determination of equilibrium constants. The Nernst Heat Theorem

(*a*) This final section concerns itself with the problem of the determination of equilibrium constants from the results of calorimetric measurements alone. For this purpose it will suffice to take the system to be a mixture of ideal gases undergoing just one chemical reaction. To begin with, one can of course measure $K(T)$ at any particular temperature by mixing specified amounts n_{0i} of the various constituents, and measuring the consequent equilibrium concentrations c_i. Then, in view of (90.5, 7), the value of $K(T)$ may be calculated, and thereafter the equilibrium concentrations will be known (at that temperature) for any choice of the n_{0i}. If the values of $K(T)$ have been obtained over a range of temperatures, the heat of reaction follows from (92.8), i.e.

$$\Lambda = -RT^2(\ln K)_{,T}. \tag{94.1}$$

The measurements just described are not of a calorimetric kind. Moreover, in practice, not only may the measurement of the concentration be difficult, but the crucial problem is often to *predict* their equilibrium values when given amounts of the constituents are mixed under specified conditions. On a simple level one might content oneself with inquiring into the value of $K(T)$ at temperature T if its value is known at some temperature T_1. $K(T_1)$ would have to be measured in the manner already described, but here one has the freedom of choice of the temperature T_1, and one can take it so high that the reaction rate is conveniently large. One so circumvents the difficulties caused by the very rapid decrease of reaction rates with temperature which commonly occurs. Thus, now, from (94.1), [2_2^*]

$$R\ln\big(K(T)/K(T_1)\big) = -\int_{T_1}^{T} \Lambda(T)\,dT/T^2, \qquad (94.2)$$

whilst from (92.7), since $h_{,T} = c_P$,

$$\Lambda(T) = \Lambda(T_1) - \int_{T_1}^{T} \sum_i \nu_i c_{Pi}\,dT. \qquad (94.3)$$

Hence $K(T)$ can be calculated if one has determined the equilibrium constant and heat of reaction at some temperature T_1, together with the specific heats of the various constituents in the temperature range between T_1 and T.

(*b*) The fully fledged problem is now this: to calculate $K(T)$ without recourse to *any* experimental results relating to the direct measurement of equilibrium concentrations. Here one often finds a somewhat bewildering wealth of detail, involving all sorts of constants of integration—the vapour pressure constants, chemical constants, and the like—which unnecessarily obscures the theory. Moreover, there is a tendency to go through a series of steps in the course of which various relations are differentiated, only to be integrated again later on. This shall be avoided here. Essentially one is confronted with the task of calculating K directly from (90.4), which defines it. Thus, in view of (82.16) and (90.2)

$$\ln K(T) = \nu \ln P - \Sigma \nu_i g_i/RT. \qquad (94.4)$$

Now $g_i = h_i - Ts_i$, so that one gets, on using (92.7),

$$R\ln K(T) = \Lambda/T + \Sigma \nu_i (s_i + R\ln P). \qquad (94.5)$$

Thus $K(T)$ can be determined calorimetrically if $\Sigma\nu_i s_i$ can be so determined. Accordingly, consider first the expression $s+R\ln P$, where s is the molar entropy of a typical constituent C. Now, quite generally, $s_{,T} = \tilde{c}_P$, the coordinates being P and T, of course. Hence, if s_0 is a constant, [13]

$$s = s_0 + \int_0^T \tilde{c}_P dT. \qquad (94.6)$$

Concerning this equation various remarks need to be made. First, the integration has been extended down to $T = 0$. Of course, C does not remain an ideal gas. On the contrary, as the temperature is lowered, the substance will undergo phase changes. However, the Third Law requires that the integral shall converge. Secondly, the singularities in c_P, associated with the phase changes, are to be understood to have been accommodated after the manner of Section 83(vii), so that the integral in (94.6) takes latent heats of transformation correctly into account. Thirdly, the integral is to be evaluated at constant pressure, so that c_P is the specific heat at constant pressure measured at the pressure P. Fourthly, if T_0 is the lowest temperature which can be attained in practice, the values of c_P for $T < T_0$ have to be obtained by extrapolation. The fifth, and last, remark concerns the constancy of s_0. This again is implied by the Third Law: s_0 being a function of P only, it can only be a constant, since $s \to s_0$ as $T \to 0$, and this must be independent of P.

Recalling (94.1), it will be noticed that the sum on the right of (94.5) cannot depend upon P. It follows that the quantity

$$e(T) = R\ln P + \int_0^T \tilde{c}_P dT \qquad (94.7)$$

is a function of T only. Then

$$R\ln K(T) = \Lambda/T + \Sigma\nu_i e_i(T) + \Sigma\nu_i s_{0i}. \qquad (94.8)$$

What can be said about the value of the constant $d = \Sigma\nu_i s_{0i}$ which has made its appearance? Contemplate the entropy S of a system made up of n_i moles of the individual condensed constituents C_i ($i = 1, \ldots, z$). At $T = 0$ the value of S is $S_0 = \Sigma n_i s_{0i}$. When the reaction characterized by the equation $\Sigma\nu_i C_i = 0$ has proceeded to an extent given by the value ξ of the degree of advancement, S_0 will have changed by an amount ξd. [Whether or not the reaction will actually occur near $T = 0$ is irrelevant, since it may be thought of as having been brought about in a manner akin to that discussed in Section 60*a*. It certainly will not occur *at* $T = 0$ since the system can reach no such state.] According to the Third Law, S_0 is, however, independent of the degree of advancement of the reaction:

$$d = 0, \tag{94.9}$$

a result concerning whose validity a brief comment will follow in a moment. Now $[2^*_3]$

$$R \ln K(T) = \Lambda/T + \Sigma \nu_i e_i(T), \tag{94.10}$$

so that $K(T)$ can be obtained by means of calorimetric measurements alone.

(*c*) As a matter of interest, (94.9) is a direct expression of the so-called *Nernst Heat Theorem*, to which, historically, the Third Law owes its origin. That it cannot be valid without certain reservations may be seen as follows. $K(T)$ and Λ are definite quantities, and the same must therefore be true also of $d + \Sigma \nu_i e_i$. The Nernst Law (94.9) declares that d is to be taken as zero, which incidentally implies that one cannot assign arbitrary values to the zero point entropies of the substances taking part in a chemical reaction. This law is now to be verified experimentally, which may be done by measuring heats of reaction, specific heats, and equilibrium concentrations. The issue at stake is adequately illustrated by supposing that at sufficiently low temperatures all constituents but the first are pure crystals, but that C_1 is the kind of substance M discussed in Section 61(*b*), so that $s'_{01} \neq s''_{01}$, the prime and double prime referring to the crystalline and amorphous form of C_1 respectively. Evidently d cannot be taken as zero independently of the form in which C_1 exists whilst measurements are carried out. Quite generally, the validity of (94.9) must therefore be restricted: it shall apply only under conditions in which the condensed constituents have a certain definite generic structure, which experiment shows to be that of a pure crystal. One can think of the vanishing of d as expressing the vanishing of the entropy change in a reaction between pure crystals as $T \to 0$; and Nernst's Theorem is often expressed in this way. The specific heats which enter into (94.8) must therefore relate to pure crystalline forms of the constituents.

Equation (94.9) appeared above as a consequence of the Third Law. Yet it has been concluded that the vanishing of d is not assured without reservation. How can this be? The explanation of the apparent contradiction is implicit in the discussion of Section 61*c*. It is to be found in the incautious use of the Third Law under conditions in which a lack of uniqueness may make it inapplicable. Clearly one must require that the coordinate-independent values of the entropy function to which the Third Law refers should be *definite* under the circumstances in which the law is used to draw conclusions about the behaviour of any particular system. When a constituent like C_1 is present this condition is not necessarily satisfied since, as $T \to 0$, C_1 need not end up as a regular crystal. The ambiguity inherent in an ill-defined, irregular structure of a constituent brings with it that the entropy of this constituent may have one of several values, or even of a whole range of values, as $T = 0$. In this sense the condition of the system is in general not properly governed by the coordinates alone. Uniqueness is restored by requiring that, for the purposes of measurement, all constituents must be pure, regular crystals at sufficiently low temperatures: in this unambiguous situation the Third Law can then be applied here without reservation.

BIBLIOGRAPHY

CALLEN, H. B. (1960). *Thermodynamics.* New York, Wiley.

GUGGENHEIM, E. A. (1950). *Thermodynamics.* Amsterdam, North-Holland.

Handbuch der Physik, Vol. III/2. Berlin, Springer. (The articles by Guggenheim, E. A.: Thermodynamics, classical and statistical; and by Falk, E. and Jung, H.: Axiomatik der Thermodynamik.)

LANDAU, L. & LIFSHITZ, E. (1938). *Statistical Physics.* Oxford University Press.

LANDSBERG, P. T. (1961). *Thermodynamics.* New York, Interscience.

PIPPARD, A. B. (1957). *Elements of Classical Thermodynamics.* Cambridge University Press.

PRIGOGINE, I. & DEFAY, R. (1954). *Chemical Thermodynamics.* London, Longmans.

TOLMAN, R. C. (1934). *Relativity, Thermodynamics and Cosmology.* Oxford University Press.

WILKS, J. (1961). *The Third Law of Thermodynamics.* Oxford University Press.

WILSON, A. H. (1957). *Thermodynamics and Statistical Thermodynamics.* Cambridge University Press.

ZEMANSKY, M. W. (1957). *Heat and Thermodynamics.* New York, McGraw-Hill.

In these volumes the reader will find alternative treatments of classical thermodynamics, more extensive or advanced treatments of specific applications, and references to the original literature, the last mentioned especially in Wilson's excellent treatise.

INDEX